Perspectives on the History of Chemistry

Series Editor

Seth C. Rasmussen, Department of Chemistry and Biochemistry, North Dakota State University, Fargo, ND, USA

Commonly described as the "central science", chemistry and the chemical arts have an extremely long history that is deeply intertwined with a wide variety of other historical subjects. Perspectives on the History of Chemistry is a book series that presents historical subjects covering all aspects of chemistry, alchemy, and chemical technology.

Potential topics might include:

- An updated account or review of an important historical topic of broad interest
- Biographies of prominent scientists, alchemists, or chemical practitioners
- Translations and/or analysis of foundational works in the development of chemical thought

The series aims to provide volumes that advance the historical knowledge of chemistry and its practice, while also remaining accessible to both scientists and formal historians of science. Volumes should thus be of broad interest to the greater chemical community, while still retaining a high level of historical scholarship. All titles should be presented with the aim of reaching a wide audience consisting of scientists, chemists, chemist-historians, and science historians.

All titles in the book series will be peer reviewed. Titles will be published as both printed books and as eBooks. Both solicited and unsolicited manuscripts are considered for publication in this series.

More information about this series at http://www.springer.com/series/16421

Brian Halton

Some Forgotten Chemists

 Springer

Brian Halton
School of Chemical and Physical Sciences
Victoria University of Wellington
Wellington, New Zealand

ISSN 2662-4591   ISSN 2662-4605 (electronic)
Perspectives on the History of Chemistry
ISBN 978-3-030-16402-7   ISBN 978-3-030-16403-4 (eBook)
https://doi.org/10.1007/978-3-030-16403-4

Library of Congress Control Number: 2019936279

This collection of essays consists of selections from the series originally published quarterly as Some Unremembered Chemists in the New Zealand Institute of Chemistry in-house journal, Chemistry in New Zealand (2013–2018). They are abstracted, edited and abbreviated slightly, and appear with the permission of the copyright holder.
© Springer Nature Switzerland AG 2020
This work is subject to copyright. All rights are reserved by the Publisher, whether the whole or part of the material is concerned, specifically the rights of translation, reprinting, reuse of illustrations, recitation, broadcasting, reproduction on microfilms or in any other physical way, and transmission or information storage and retrieval, electronic adaptation, computer software, or by similar or dissimilar methodology now known or hereafter developed.
The use of general descriptive names, registered names, trademarks, service marks, etc. in this publication does not imply, even in the absence of a specific statement, that such names are exempt from the relevant protective laws and regulations and therefore free for general use.
The publisher, the authors and the editors are safe to assume that the advice and information in this book are believed to be true and accurate at the date of publication. Neither the publisher nor the authors or the editors give a warranty, expressed or implied, with respect to the material contained herein or for any errors or omissions that may have been made. The publisher remains neutral with regard to jurisdictional claims in published maps and institutional affiliations.

This Springer imprint is published by the registered company Springer Nature Switzerland AG
The registered company address is: Gewerbestrasse 11, 6330 Cham, Switzerland

Acknowledgements

Butlerov, et. al: I am especially grateful to Prof. Vladimir Vysotskii for his invaluable comments and many helpful suggestions for this article. Professor Andrei Vedernikov (University of Maryland) has been of special assistance in gaining permission to reproduce the image of Butlerov and his colleagues (Fig. 6.3) from Prof. Vladimir Ivanovich Galkinand (Butlerov Chemical Institute) and Prof. Andrei Vedernikov for providing the caption for it.
    Baekeland: I am grateful to Reindert Groot of the Amsterdam Baekeland Collection for helpful corrections and comments.
    Easterfield: I am grateful to Richard Rendle for suggesting the topic and providing Easterfield's New Zealand Institute of Chemistry's application form, and to Andrea Mead of the Cawthron Institute, Nelson, New Zealand, for assistance and the provision of much useful information.
    Mellor: I am grateful to Dr. Ian Brown for comments on the manuscript and for providing the NZ records on Alfred Mellor.
    Mercer: I am grateful to Dr. Mattie Timmer for improving the structures depicted in Figs. 13.3 and 13.4.
    Richards: I am grateful to Drs. Joanne Harvey and Margaret Halton for helpful comments.
    Women Pioneers: I thank Prof. Jay Siegel for providing a copy of ref. 1 and Dr. Irène Studer-Rohr, Public Relations, Chemistry Department, University of Zurich, for the copy of Sesemanns' official university record.
    I would especially like to thank my compositer and the publisher of *Chemistry in New Zealand*, Rebecca Hurrell, for her unfailing attention to the detail of the text of this book. Her support and friendship over the years of my formal retirement and the time I was involved with the journal have made life so much better than it would otherwise have been. Finally, I thank Seth Rasmussen for his significant editorial input and support. Springer has generously transformed the text to the book that it now is.

Contents

1	**Introduction**	1
	References	2
2	**Carl Friedrich Accum (1769–1838)**	3
	References	12
3	**Henry Edward Armstrong (1848–1937)**	15
	References	22
4	**Leo Henricus Arthur Baekeland (1863–1944)**	25
	References	32
5	**Alexander Porfirevich Borodin (1834–1887)**	35
	References	45
6	**Aleksandr Mikhailovich Butlerov (1828–1886) and the Cradle of Russian Organic Chemistry**	47
	6.1 Brief Commentaries on Butlerov's Noted Students in the Cradle of Russian Organic Chemistry	55
	6.1.1 Vladimir Vassil'evich Markovnikov (1838–1904)	55
	6.1.2 Alexandr Nikiforovich Popov (1840–1881)	55
	6.1.3 Aleksandr Mikhaylovich Zaitsev (1841–1910)	55
	6.1.4 Ivan Laverntievich Kondakov (1857–1931)	56
	6.1.5 Alexey Yevgrafovich Favorskii (1860–1945)	56
	References	56
7	**Heinrich Caro (1834–1910)**	59
	References	69
8	**John William Draper (1811–1882)**	71
	References	79

9	Sir Thomas Hill Easterfield (1866–1949)	81
	References	89
10	Sir Edward Frankland (1825–1899)	91
	References	98
11	William Henry (1774–1836)	99
	References	105
12	Joseph William Mellor (1869–1938)	107
	References	115
13	John Mercer (1791–1866)	117
	References	127
14	Ellen Swallow Richards (1842–1911)	129
	References	137
15	Philip Wilfred Robertson (1884–1969)	139
	References	146
16	Women Pioneers	147
	References	156
17	William John Young (1878–1942)	159
	References	162

About the Author

Prof. Brian Halton (1941–2019) was educated at the University of Southampton, UK, graduating with a B.Sc. (Hons) in 1963 and a Ph.D. in 1966, with studies for the latter being under the guidance of Richard Cookson. He then moved to the University of Florida as a postdoctoral fellow with Merle Battiste, before being engaged briefly as an assistant professor. His subsequent relocation to Wellington, New Zealand, in 1968 ended up being a permanent one. He started at Victoria University of Wellington (VUW) as a lecturer and advanced through the promotion levels to professor in 1991.

A major focus of professor Halton's research was the chemistry of cyclopropanes—pungent smelling, strained organic molecules containing a cyclopropane ring fused to various aromatic rings. He published over 120 peer-reviewed original research articles on this subject and supervised 11 Ph.D. students and 36 honours/masters students, as well as several postdoctoral and other research employees. Brian Halton's contributions to chemistry through his research, teaching, writing and service have been recognised through numerous awards and fellowships. He was made an Honorary Fellow of the New Zealand Institute of Chemistry (2005) and a Fellow of the Royal Society of New Zealand (1992). Among other accolades, Brian was awarded the ICI Medal for Excellence in Chemical Research (1980), a Fulbright Award (1981), the New Zealand Association of Scientists' Shorland Medal (2001) and gave the 2003 NZIC Wellington Branch Mellor Lecture. The New Zealand Institute of Chemistry (NZIC) has instituted the triennial Wellington Branch Halton Lecture, for which the first two lecturers were Martin Banwell (ANU, 2014) and David Officer (Wollongong, 2017).

Brian Halton contributed enormously to New Zealand science through his decade-long tenure on the Pacifichem organising committee and many roles in the NZIC. He was the NZIC President during 1986–1987, the Editor of Chemistry in New Zealand for a decade (2001–2011) and then the Consulting Editor until 2018. Locally, Brian was the Wellington Branch Editor for over 15 years and on the local committee for over a decade.

Brian Halton was editor of the book series Advances in Strain in Organic Chemistry (JAI Press) for a decade. He co-wrote the 1974 textbook Organic Photochemistry (Cambridge University Press) with Jim Coxon (Canterbury), which enjoyed an expanded second edition (published in 1987), and is still found on researchers' bookshelves the world over. His recent books on his life and the history of chemistry at VUW, From Coronation Street to a Consummate Chemist and Chemistry at Victoria: the Wellington University, are both highly informative, written with his astute sense of humour and available at https://www.victoria.ac.nz/scps/about/history/history-of-chemistry, as well as in print. He also published an e-book on his encounters with heart disease, intended as a resource for others with similar conditions, entitled A Cat of Nine Lives – and the Beat Goes On: Living with heart disease, available at https://archive.org/details/ACatOfNineLives-AndTheBeatGoesOn_249. Post-retirement, Brian continued to contribute to Chemistry in New Zealand, the journal of the local chemistry professional body, with a Today in History column that ran for several years, his Unremembered Chemists series (2013–2018), and a piece to mark the International Year of the Periodic Table in the first 2019 issue.

Chapter 1
Introduction

Each of the chapters in this compilation of 17, save this one, covers the lives and work of selected chemists who have made a significant contribution to the advancement of the discipline, the profession, and well-being of humankind, yet who are little remembered. They are reproduced with the permission of the New Zealand Institute of Chemistry from the 24 articles that appeared in *Chemistry in New Zealand*, its journal, between 2013 and 2018 under the title of *Some Unremembered Chemists* [1–17]. The chapters that appear have been carefully chosen to fit to the book series with each being further edited, refined, and modified as needed. Chapter 17 on W. J. Young has been abstracted from that on Sir Arthur Harden and William John Young [17] to better fit to the book size of the series. Throughout, the images and chemical structures shown are realigned to fit to the *Perspectives on the History of Chemistry* format. The essays appear in alphabetical order so that the reader is able to read them in the order that suits best as each chapter is complete in itself, including the reference citations and notes.

References

1. Halton B (2017) Carl Friedrich Accum (1769-1838). Chem N Z 81:172–178
2. Halton B (2015) Henry Edward Armstrong, FRS (1848–1937). Chem N Z 79:157–160
3. Halton B (2016) Leo Hendrick Baekeland (1863-1944). Chem N Z 80:148–153
4. Halton B (2014) Alexander Porfirevich Borodin (1834–1887). Chem N Z 78:41–47
5. Halton B (2018) Aleksandr Mikhailovich Butlerov (1828–1886) and the Cradle of Russian Organic Chemistry. Chem N Z 82:46–52
6. Halton B (2015) Heinrich Caro (1834–1910). Chem N Z 79:200–206
7. Halton B (2013) John William Draper (1811–1882). Chem N Z 77:136–141
8. Halton B (2017) Sir Thomas Hill Easterfield (1866-1949). Chem N Z 81:83–88
9. Halton B (2013) Sir Edward Frankland KCB, FRS, FCS (1825–1899). Chem N Z 77:92–96
10. Halton B (2014) William Henry, MD, FRS (1774-1836). Chem N Z 78:128–131
11. Halton B (2014) Joseph William Mellor, CBE, FRS (1869–1938). Chem N Z 78:85–89
12. Halton B (2013) John Mercer FRS, FCS, MPhS (1791-1866). Chem N Z 77:22–26
13. Halton B (2013) John Mercer Part II. The industrialist, the chemist and the man. Chem N Z 77:52–55
14. Halton B (2016) Ellen Swallow Richards (1842–1911). Chem N Z 80:195–201
15. Halton B (2015) Philip Wilfred Robertson (1884–1969). Chem N Z 79:51–55
16. Halton B (2018) Women Pioneers in Chemistry. Chem N Z 82:192–197
17. Halton B (2014) Sir Arthur Harden, FRS (1865-1940) & William John Young (1878–1942). Chem N Z 78:169–174

Chapter 2
Carl Friedrich Accum (1769–1838)

Friedrich Accum (Fig. 2.1) is one of those of whom it has been said '[he] is representative of a chemist who is largely forgotten, but nevertheless contributed to important changes in our society' [1]. He was an apothecary (a person who prepared and sold medicines or drugs), who then established a chemicals and equipment business in early 19th century London, where he provided popular practical and theoretical instruction in chemistry. He became interested in gas lighting and was intimately involved with the design of London's first gasworks. Most importantly, Accum was the first to publicise the need for protecting the quality of foodstuffs with his popular book *The Adulterations of Foods and Culinary Poisons* [2] that sold 1000 copies within a month of its 1820 publication. It was reprinted and ran to a second edition in the same year, with a third the following year. It was published in Philadelphia (1820), and it had a German translation published in Leipzig in 1822.

Friedrich Christian Accum was born on March 29, 1769 in Bückeburg, a former capital of the tiny principality of Schaumburg-Lippe in Lower Saxony some 30 km southwest of Hanover in Germany. His father, Markus Herz, was of Jewish descent and born in nearby Vlotho on the Weser River, who had served as a carbineer in Count Wilhelm von Shaumburg-Lippe's infantry. In 1755, Markus converted from Judaism to Protestant Christianity, changing his name to Christian Accum (Accum or Akum: non-Jewish). Within a few months of his conversion, Christian married Judith Susanne Marthe Bert La Motte, the daughter of a hat maker whose Huguenot family had also suffered persecution, and with whose hand came with an ancestral home at 141 School Street in Bückeburg [3, 4]. Accum senior used the home as his business premises as a merchant and soap boiler and became moderately prosperous. The couple had seven children born in that home with Friedrich being the sixth. Sadly, only he, his elder brother and a sister (Wilhelmina) survived to adulthood. Christian Accum died in 1772, soon after his last child was born and when Friedrich was merely

---

A previous version of this chapter was published as Halton B (2017) Carl Friedrich Accum (1769–1838). Chem N Z 81:172–178.

© Springer Nature Switzerland AG 2020
B. Halton, *Some Forgotten Chemists*, Perspectives on the History of Chemistry,
https://doi.org/10.1007/978-3-030-16403-4_2

Fig. 2.1 Fredrick Accum from the *European Magazine* (July 1820) (engraving by James Thomson from Wikimedia)

three-years old. It was his older brother Philip who took over the family business and perhaps instilled an interest in chemistry in his sibling.

Friedrich attended the Bückeburg Gymnasium. The curriculum was classical and he received essentially no scientific education until after he had graduated. Shortly afterwards he became associated with the Brande family, who were apothecaries to George III, King of Hanover and Great Britain, and family friends. Friedrich trained as an apothecary prior to moving to London in 1793 where he was an assistant at the Brande pharmacy in Arlington Street (off Piccadilly). He was classified as a chemist in the register of the Alien Office. The pharmacy proprietor was the father of noted chemist William Thomas Brande, but the buisness was run by his elder brother Everard. Accum remained at the pharmacy until about the turn of the century gaining knowledge and friends, one of whom was the gifted chemist William Nicholson. He assisted Nicholson with his writings. Accum also became one of the most frequent early contributors to the *Journal of Natural Philosophy*, subsequently better known as *Nicholson's Journal*, which first appeared in 1797. From 1798 to 1803 Fredrick had seven contributions on a variety of topics [5]. However, his second paper [6], an attempt to discover the genuineness and purity of drugs and medicinal preparations, marked his first foray into erroneous analysis and misrepresentation, and the beginning of his food technology and analytical expertise [3]. The September and October issues of Nicholson's 1799 *Journal* carried a translation of Achard's report on his sugar production from local German sugar beet. Accum took it upon himself to obtain a sample for Nicholson, whose analyses led him to the conclusion that there was little difference between the beet sugar and loaf-sugar. This sugar beet sample was the first admitted to England and marked the beginning of importation that reached 70,000 tons in 1900 [3].

During the 1790s, Accum anglicised his given name from Friedrich to Fredrick. He married Mary Ann Simpson in 1798. They had eight children, but only two survived birth or infancy, their eldest child Flora Eliza (b. May 17, 1799) and Friedrich Ernst (b. April 3, 1801) [4]. In 1800, they moved to 11 Old Compton Street. It was there that Fredrick set up his laboratory and business for the analysis and examination of commercial products, and supply of chemicals and apparatus. Once on a firm footing he provided instruction in the theory and practice of chemistry. His activities described on his business card were as follows [5]:

> Mr Accum acquaints the Patrons and Amateurs of Chemistry that he continues to give private Courses of Lectures on operative and Philosophical Chemistry, Practical Pharmacy and the Arts of Analysis, as well as to take resident Pupils in his House and that he keeps constantly on sale in as pure a state as possible all the Re-Agents and Articles of research made use of in Experimental Chemistry, together with a Complete Collection of Chemical Apparatus and Instruments calculated to Suit the convenience of Different Purchasers. Philosophical Gentlemen residing in the Country or Abroad desirous of becoming purchasers of large or small Collections of Chemical Preparations, etc., may have explanatory Lists previously made out to the Expense they are willing to incur, and Chemical catalogues may be had at the laboratory, Old Compton Street, Soho, London.

For many years, Accum's establishment was the only one to offer lectures on the theory of chemistry, with practical training in a laboratory. Amateurs were welcome to perform simple experiments on site. His teachings attracted various prominent students. These included Lord Palmerston who later became England's Prime Minister, and then the first American scientists of whom Benjamin Silliman became Professor of Chemistry at Yale College, giving the first chemistry lectures there in 1804 and furnishing his laboratory with equipment from Accum. Another was William Dandridge Peck, a botanist and subsequent Professor at Harvard.

In 1801, Accum was appointed as Assistant Chemical Operator at the Royal In-stitution (RI) when Humphrey Davy was Director there. His skill and ability in constructing apparatus had become known before this appointment and it gained further recognition from those who attended the Institution lectures. The appointment lasted for two years as Accum resigned shortly after his supporter, Count Rumford, left England for Paris to marry Antoine Lavoisier's widow [5]. In 1803, before he left, Accum published his two volume *System of Theoretical and Practical Chemistry* [7]. Its importance lies in the fact that this was the first textbook of general chemistry to be written in English, based on Lavoisier's principles. It was much expanded and improved for the 1807 second edition.

It was increasingly felt that there was room for another establishment, separate from the RI, which would ally its commercial and manufacturing concerns more closely. By 1808 the Surrey Institution had been established for this purpose. It was located in the former home (1789–1806) of the Leverian Museum at 3 Blackfriars Road, on the south side of Blackfriars Bridge, and remained there until its closure in 1823. Its lecture theatre was known as the Rotunda. Here, Fredrick Accum initiated

Fig. 2.2 Left: Chemical Lectures, contemporary caricature by Thomas Rowlandson, after 1809. The cartoon has Accum lecturing and (standing on his right) R. Ackermann (publisher and Accum's and Rowlandson's friend), and Sir J. Hippisley (Manager-Royal Institution) and (an on his left) Count Rumford (sitting with his back to the audience), and (possibly) Sir Humphry Davy standing between the columns (photographic reproduction of a two-dimensional, public domain work of art from wikmedia Commons). Right: The first London gasworks, 1814. Plate from Accum's A Practical Treatise on Gas Light (1815) (Project Guttenberg). The retorts are set transversely, directly under the chimney; the gasometer is on the left

the chemistry programme with the Institution's second course, his January 1809 lectures on Chemistry and Mineralogy. Chemistry of the Arts (1811) and Chemical Phenomena of Nature and Art (1812) followed this. He returned to provide the chemistry course in 1819 (Chemistry Applied to the Arts and Manufacture) and 1820 (Chemical Phenomena in Nature and Art). The Rotunda was caricatured in 1808 by Thomas Rowlandson and Charles Pugin, and with Accum lecturing there in 1810 by Rowlandson (Fig. 2.2) [8]. There was debate on the identity of this lecturer with some [2, 5] suggesting that it could be Davy. However, the history of the Surrey Institution by Kurzer [8] makes it clear that Davy never lectured there and that the cartoon is of Accum with dignitaries from the RI in attendance and that it is from 1810.

The Surrey Institution became a significant centre in the intellectual life of London, with its lecture theatre being one of the most elegant institution rooms in the city. It contained two galleries within its diameter of about 12 metres, with the ground floor holding nine rows of seats that gradually rose above each other. Seating was for up to 500 in total. The courses on chemistry were one of the mainstays of the Institution programme. In those days, chemistry was regarded as the foremost of all the sciences and an essential adjunct to all others by providing rational explanations of familiar and obscure phenomena. It was of the utmost importance to manufacture and commerce, and its relevance to medicine and pharmacy was equally obvious. Thus, a course on chemistry was an invariable component of the season's activity at the Surrey Institution. It was the most systematic course and usually the longest, with the greatest content of all the programmes offered. That Accum gave the first courses stemmed from his expertise in the analysis of ores and minerals, the detection of adulteration of drugs and medicinals and his experience in conducting courses in chemistry in his own laboratory.

Accum's lectures on experimental chemistry and analytical mineralogy actually commenced at the Chemical Laboratory, Compton Street, Soho, on October 18, 1808. They comprised the practical operations of the scientific laboratory, general rules to be observed in the performance of experiments, and experimental elucidations in the science of chemical philosophy and its application to the useful arts. Accum's remit was to provide instruction for initiating into the principles of chemical philosophy those who possess no previous knowledge of it. While these principles remained the core of his teaching, he changed the emphasis from year to year, widening the coverage of the subject and sustaining the interest of a regular audience. His successive courses highlighted the relationship of chemistry with mineralogy, metallurgy, natural phenomena, and application to the manufacture and the arts.[1] To assist his audience, Accum issued a booklet listing the heads of topics of his 17 lectures. His first course dealt with minerals, ores, metals, and their analysis. In 1810, he published a more elaborate companion to his subsequent lecture courses, providing a permanent record of his chemical teaching.

It was during the 1800–1810 period that Accum became fascinated with coal gas production [3, 5]. He appears to have attended the first demonstrations of gas lighting in London by the German technologist, H. A. Winsor. Winsor illuminated the Lyceum Theatre during the 1803–1804 winter and then gained a patent for illuminating gas production in 1804. He gleaned promotors for a company to generate and use coal gas. Winsor went on to demonstrate that gas lighting was practical after he had purchased a home on Pall Mall. He piped gas from his home to street lights that he had erected in the Mall to provide the first public street lighting [9] on January 28, 1807. By then, Accum had shown interest in gas production and had been approached by the promoters of Winsor's proposed company. He conducted a long series of experiments to provide supporting scientific evidence. These and the data from them formed the basis of his chemical expert testimony before the Commons and then the Lords Committees in 1809 [3]. Accum paid particular attention to the by-products of gas manufacture and provided samples of them. Interestingly, theses samples were held at the Old Ashmolean Museum in Oxford in 1938 and, as Browne states [4], it would be very interesting to have that labelled Highly Rectified Essential Oil analysed to see how closely it compares to Faraday's first sample of benzene discovered in 1825, 16 years later.

The Commons and Lords committees finally gave Winsor the needed approval for a Chartered Gas-Light and Coke Company in 1810, but it stipulated that it must raise £100,000 before the company could be given its charter. This happened in 1812 and the Gas Light and Coke Company was incorporated with Accum as its chemist and Board member. The charter states that the company was to provide *inflammable air for the lighting of the streets, oil, tar, pitch, asphaltum, ammoniacal liqor, and essential oil from coal and for other purposes relating thereto.* Accum planned and oversaw the construction of the gas plant on Curtain Road, the first such plant in

---
[1] See the appendix in Ref. [8].

the history of gaslight (Fig. 2.2). On December 31, 1813 Westminster Bridge was lit by gas and in the City of London gas lighting began on Christmas Day in 1814 [10]. Accum resigned from the company in November 1813 [5] though he went on to install a number of other plants. He was the recognised expert and had some 15-year association with coal-gas technology. His publications on the subject provided all the detail needed to build a plant, even to the extent (noted in his book) of giving the cost of having Accum install the plant (1st edn.: £1940-11-0) [11, 12]. The rapid growth of gas works in London led to sewer and river pollution from discharge of the sulfur and tar by-products. Accum demanded legal measures to prevent the discharge of these materials but his criticisms met with little support; the various gas explosions that happened from mismanagement had more impact [1].

Fredrick Accum suffered financially following the end of the Napoleonic wars in 1814. Nevertheless, he was much in demand as an expert witness as illustrated in his role for the Severn, King & Co., sugar refiners [3]. A fire almost destroyed the refinery in 1819, some three months after oil heating was installed to boil the sugar. This heating led to less caramelisation when the oil was heated to 180–200 °C. The Imperial Insurance Co., refused to meet the company claim because, in its view, the traditional open fire was safer. Accum provided chemical evidence and testified that had there been an explosion in the oil gas in the fill house, as claimed by the defendants, the noise of the explosion would have been heard throughout London. Michael Faraday appeared for the insurers but the experiments he had conducted and described as evidence were shown not to comply with the actual conditions. The jury found in favour of the company and Faraday refused to serve as an expert witness ever again.

In addition to operating his supply house, teaching, industrial gas working, and being a prolific author, Accum provided various preparations for industrial purposes. Of these, his compound for welding iron is notable [3]. Here he found combining flowers of sulfur (1 oz) with sal ammoniac (NH_4Cl, 2 oz) and iron dust (16 oz) gave the best mix. For welding, one part of this mixture was to be ground in a mortar with 20 parts of fine iron filings and water added to give a paste of good consistency. The paste was then smeared on the broken joint. Chemically, iron sulfide is produced and generates sufficient heat to weld the broken joint; it was extensively used in its time. Accum also offered public service. The small town of Thetford northeast of Cambridge engaged him to analyse their mineral spring and comment on the merits of its waters, a topic which he had studied from 1808 and published extensively on. This he did, providing a most satisfactory report on the chemical and medicinal properties of the water that led to commissioning him to provide plans for a bathing establishment. It opened in 1820, a year after his book *Guide to the Chalybeate Spring of Thetford* [13] was published; it was 'for the convenience of the public'.

Despite his significant reputation, Fredrick Accum went from saint to sinner within just a few months of the publication of his most famous work *The Adulterations of Foods and Culinary Poisons* [2]. As mentioned earlier, in 1798 he published an article in Nicholson's Journal on drug and medicinal purity [6] that formally marked the first chemical attempt to raise matters of drug (and food) safety by drawing attention to the prevalence of adulteration and the dexterity with which it was practised. It was the continuation of this as occurred in foods that led to his downfall in England and his subsequent return to Germany. In January 1820, his book on food adulteration appeared and gained immediate popular acclaim [2]. It is a classic, in that it was the first such book to deal with the subject and the difficult problems of its subject matter. It was from his experiences, beginning in 1797, as an analyst and food technologist that gave him a knowledge of the subject far exceeding that of any other chemist of the time. He exposed practices that led to enormous comment and aroused feelings among those who were involved in adulteration. It is thought [3] that the text is the most extensively reviewed book on chemistry ever written. In his essay, Charles Browne provides two and a half pages of small print reproductions of reviews, and these are only from the first edition of the book, which also drew attention in the US. Accum's downfall stems from his listing at the end of each chapter the names of those who had been guilty of adulteration. He made the comment that *the man who robs a fellow subject of a few shillings on the highway is sentenced to death; while he who distributes a slow poison to a whole community escapes unpunished*. However, he did state in his Preface that he found naming people and companies an invidious office and a painful duty. Moreover, he only published the names of those who had been authenticated in Paliamentary documents and other records.[2] His book is written in a straightforward manner with description of simple analytical tests provided for his readers. It was at a time when more central industrial manufacture and the availability of additives (chemicals) were increasingly employed. The proliferation of newly discovered chemicals and the absence of laws controlling their usage meant that unscrupulous merchants could adulterate foods boosting their profits at the expense of public health.

The book's content alternates between harmless forgeries such as the mixing of dried pea grounds in coffee and the use of markedly more dangerous additives [1]. As an example, the high lead content of Spanish olive oil came from lead containers used to clear the oil, so Accum recommended the use of oil from countries where this was not the practice. Vinegar was often contaminated by the addition of sulfuric acid to increase the acidity. Many green sweets sold by merchants on the streets of London were adulterated with sap green (from buckthorn berries) that have a high copper content. James Millar in a letter dated September 22, 1819 to the *Philosophical Magazine* drew attention to a charwoman whose green tea was adulterated with copper as proved by Accum's analyses [14]; this is described in detail by Cole [5]. Particular attention was paid to beers. English beer was sometimes contaminated with molasses, honey, vitriol, pepper and occasionally opium. One of the worst contaminations was the addition of fish berries (from the Menispermaceae family) to port as they contain picrotoxin. His analyses were supported with evidence from the quantity and time of imports and the consequential variation in the cost of the berries. Accum's book cover and its frontispiece, was provoking. He designed most of his book covers himself, none more imaginative than his book on food adulteration. The

---

[2]See Ref. [2], p. v.

cover depicts a spider web with the spider hovering over its victim, the whole encased by 12 serpents with forked tongues and intertwined tails, while the frontispiece carries a pot holding a death's head with the inscription: *There is death in the pot*, a name by which the book became known.

Noel Coley in his article [15] *The Fight against Food Adulteration* comments on Accum's work stating that by the early 19th century "... tea and coffee drinking had become popular in England but, being imported, both were expensive and as the fashion spread, cheaper varieties were needed for sale to the masses". Many of these were not genuine tea and coffee but were made to look like the real thing by chem-ical treatment. Spent tealeaves and coffee grounds could be bought for a few pence per pound from London hotels and coffee shops. The used tea leaves were boiled with copperas (ferrous sulfate) and sheep's dung, then coloured with Prussian blue (ferric ferrocyanide), verdigris (basic copper acetate), logwood, tannin or carbon black, before being resold. Some varieties of cheap teas contained, or were made entirely from, the dried leaves of other plants. Exhausted coffee grounds were treated in a similar way, adulterated with other roasted beans, sand and/or gravel, and mixed with chicory, the dried root of wild endive, a plant of the dandelion family. Chicory itself was sometimes adulterated with roasted carrots or turnips and the dark brown coffee colour was achieved by using 'black jack' (burnt sugar).

It is not surprising, therefore, that Accum's 1820 publication drew much praise from the public as evidenced by the appearance of a reprint and the second edition in that same year. But it also drew its critics, especially from those practicing adulteration. Subsequent events led to Accum being discredited. A few months after publication of the Adulteration of Food book, one of the library staff at the RI, assistant librarian Mr. Sturt, took to the managers a complaint against Accum claiming that he had torn pages out of the books he had been reading. Although there have been many commentaries of this and subsequent incidents [1, 3–5], that of Cole [5] reproduces various minutes of the Managers' Meetings of the RI. From these it is evident that Sturt saw Accum in the library and that pages of books and journals were removed. This was and still is done by library users today, but in the early 19th century it was more common than now as scrap paper was not as readily available then. Sturt was asked to drill a hole in a partition to watch Accum more closely. This he did and on the evening of December 20, 1820, he watched him closely for some two hours. He saw Accum remove pages from *Nicholson's Journal* and an account was passed to the RI secretary the next morning. This led to Mr. Birnie, the sitting magistrate at Bow Street, issuing a warrant to search Accum's house. Leaves from RI books were found and Accum was prosecuted and brought to trial. After hearing all the evidence, the magistrate who tried the case delivered the opinion that *however valuable the books might be from which the leaves found at Mr. Accum's house had been taken, the leaves separated from them were only waste paper.* If they had weighed one pound he would have committed Mr. Accum for the value of a pound of waste paper, but as they did not, he discharged him.

This was not the end of the matter as the managers of the RI chose to have a Bill of Indictment drawn up against Accum for the offence at the next Westminster Sessions. A supporter of Accum (likely Anthony Carlisle [5]) wrote to Earl Spencer, the President of the RI seeking the proceedings to be stopped. This did not happen and trial was set for April Sessions. Accum and his two sureties appeared in court and were set recognisances of £200 and £100, respectively. When the April Sessions came and the Accum trial was called, Accum could not be found; he had returned to Germany, his reputation in England in shreds. The bail was forfeited when it became obvious he was not to return to England. Whether the enemies that Fredrick Accum made from his food adulteration book were behind the forceful RI indictment or not remains unclear, but seems likely. The remainder of Accum's career was worked out teaching in Germany.

At time of his return to Germany in 1821, Accum was 52 years old and a widower who had spent close on 30 years in England. On his return he went to the town of Athaldensleben, near Magdeburg, where his friend, Johann Nathusius, had established a factory for tobacco production in 1787 and a factory for sugar production from sugar beet between 1813 and 1816. Nathusius had an extensive library, which Accum used, but he also enjoyed his enforced freedom and the hospitality it offered. After some time a joint post as Professor of Technical Chemistry and Mineralogy at the Royal Industrial Institute (the Gewerbeinstitut) and Professor of Physics, Chemistry, and Mineralogy at the Royal Academy of Construction (the Bauakademie) was made, and in 1822 Fredrick accepted it. The latter academy saw the publication *Physiche und chemische Beschaffenheit der Baumäterialen, deren Wahl, Verhalten und zweckmassige Anwendung* that appeared in two volumes published in Berlin in 1826. It was the only publication written in German by Accum and the last of his offerings.

A few years after settling in Berlin, Accum had a house built at 16 Marienstrasse (later No. 21) where he lived until his death. He suffered from gout during the last years of his life and, after taking a turn for the worse in June 1838, he died aged 69 years on the 28th of that month. His wife Mary Ann had died in London on March 1, 1816.

Accum's publications were extensive and include numerous papers and some 16 books [1]. Of the books not discussed above, his *Chemical Amusements, a Series of Curious and Instructive Experiments in Chemistry Which Are easily Performed and Unattended by Danger, A Treatise On The Art Of Making Wine From Native Fruits, A Treatise On The Art Of Brewing, Culinary Chemistry*, and *Elements of Crystallography: After the Method of Hauy*, illustrate the breadth and depth of his writings.

Some 12 years after Accum's death, Sir Charles Wood, the British Chancellor of the Exchequer attested that there was no chemical test to prove that coffee could be adulterated with chicory. However, Arthur Hill Hassall, a medical practitioner, microscopist and chemist, knew the statement to be untrue. He bought samples of coffees in London and examined them microscopically showing that the chicory in the coffee was easily distinguishable. He took it upon himself to prove that such adulteration was common. Between 1851 and 1854, as analyst for the new Analytical Sanitary Commission, he assessed some 2500 food and drink samples (microscopically and by chemical analysis) and showed the presence of alum in bread, iron, lead or mercury in cayenne pepper, copper salts in bottled fruits and pickles, iron oxide in sauces, and that there was 1 part of tumeric powder in 547 parts of mustard. These studies appeared in *The Lancet* as anonymous reports from the commission, only to be subsequently published by Hassall under his own name [16, 17]. This accelerated the moves for reform with *The Times* of July 24, 1855 stating in its Editorial:

> Some 30 years ago the British Public was frightened by the cry of 'Death in the Pot;' but we might now, it seems, re-echo the alarm with greater force than ever. Death is not only in the pot, it is everywhere; not only in our food and drink, but in the very medicines that should cure our diseases. The matter is now underinvestigation before a Parliamentary Committee, and it has been shown by evidence of the most convincing kind that of the articles of daily use and first necessity a very great portion is subjected to foul and systematic adulteration. But how, the reader may ask, has the discovery at this particular period been made or certified? Partly through material improvements effected in the means of detection, but mainly by the skill and perseverance of Dr Hassall, who, by devoting to this subject the energies of a scientific mind, and pursuing it with that steady zeal that its importance justified, has thus become a public benefactor of no common order.

The first Food Adulteration Act was passed in London in 1860. Subsequent concerns led to the privileged position that food safety holds today.

References

1. FamPeople, Friedrich Accum: biography. http://www.fampeople.com/cat-friedrich-accum. Accessed 04 Apr 2017
2. Accum F (1820) A treatise on adulterations of food, and culinary poisons, exhibiting the fraudulent sophistications of bread, beer, wine, spirituous liquors, tea, coffee, cream, confectionary, vinegar, mustard, pepper, cheese, olive oil, pickles, and other articles employed in the domestic economy and methods of detecting them. Longman, London
3. Browne CA (1925) The life and chemical services of Fredrick Accum. J Chem Educ 2:829–851:1008–1035, 1140–1149
4. Browne CA (1948) Recently acquired information concerning Fredrick Accum, 1769–1838. Chymia 1:1–9
5. Cole RJ (1951) Friedrich Accum (1769–1838). A biographical study. Annals Sci 7(2):128–143
6. Accum F (1798) An Attempt to discover the genuineness and purity of drugs and medical preparations. Nicholson's J 2:118–122
7. Accum FC (1803) System of theoretical and practical chemistry. Old Compton Street, Soho, London
8. Kurzer F (2000) A history of the surrey institution. Annals Sci 57:109–141
9. History House. The first demonstration of street lighting using gas. http://www.historyhouse.co.uk/articles/gas_lighting.html. Accessed 8 April 2017
10. Dictionary of Victorian London. Gas-lighting. http://www.victorianlondon.org/lighting/history.htm. Accessed 13 Apr 2017
11. Accum F (1815) Illuminating the streets by coal gas. Annals Philos 6:16–19
12. Accum F (1815) A practical treatise on gas light. R. Ackermann, London, pp v (i) 186
13. Guide to the Chalybeate Spring of Thetford. http://archive.org/stream/guidetochalybea00accugoog/guidetochalybea00accugoog_djvu.txt. Accessed 12 Apr 2017
14. Millar J (1819) On poisonous tea-leaves. Philos Mag 54:218–219
15. Coley N (2005) The fight against food adulteration. Educ Chem https://eic.rsc.org/fea-ture/the-fight-against-food-adulteration/2020253.article. Accessed 16 Jan 2019

References

16. Hassall AH (1855) Food and its adulterations; comprising the reports of the analytical sanitary commission of 'The Lancet' for the years 1851 to 1854, Longman, Brown, Green, and Longmans, London
17. Hassall AH (1876) Food: its adulterations, and the methods for their detection. Scribner & Welford, London

Chapter 3
Henry Edward Armstrong (1848–1937)

Henry Edward Armstrong (Fig. 3.1) was born on May 6, 1848 in Lewisham, then a part of Kent but now South-East London [1]. He was the first of seven children to Richard and Mary (née Biddle) Armstrong. His parents had eloped and married before they were 21 years of age in 1847 and Henry is cited as a 'seven-months child' [2]. The family remained in Lewisham for the rest of their lives. Henry also lived there his entire life, except when overseas [1, 2]. The mid-nineteenth century had Lewisham as a country village beginning to provide homes for London businessmen who liked the country air. Although the Armstrong home was always there, they moved houses as Richard's fortunes rose and fell from his position as a commission agent and importer in Mark Lane in the city of London. Thus, Richard Armstrong was a businessman with sufficient income to support his wife and increasing number of children in fair comfort. Sadly, only four of the seven children survived childhood. Henry's home in Lewisham was on Avenue Road, a street that now has disappeared and is replaced by what is the main entrance to The Lewisham Centre.

After attending a number of small schools, Henry moved to Colfe's Grammar School on Lewisham Hill bordering on Greenwich, to which he walked from home. It remains one of the oldest established schools in London. He left there at age 16 years with 'no particular interest, but observant and an experimentalist' and was thought to be 'delicate' [1]. Because of this, he spent the following winter in Gibraltar with his uncle, likely on the recommendation of his grandfather who had been Governor of the Convict Prison there. He returned to Lewisham in the spring of 1865 and 'just slid into chemistry' as his father let him attend the Royal School of Chemistry (RSC) on Oxford Street [1]. That was for the summer term of 1865 just before the noted

---

A previous version of this chapter was published as Halton B (2015) Henry Edward Armstrong, FRS (1848–1937). Chem N Z 79:157–160.

© Springer Nature Switzerland AG 2020
B. Halton, *Some Forgotten Chemists*, Perspectives on the History of Chemistry, https://doi.org/10.1007/978-3-030-16403-4_3

Fig. 3.1 H. E. Armstrong in later life from Ref. [2]. *Source* unknown

August Wilhelm von Hofmann had returned to Berlin; Henry was taught by Edward Frankland.[1] Chemistry was the only subject available at the RSC and so Armstrong took other courses at the affiliated Royal School of Mines given by Tyndall, Huxley and Ramsay. Keeble [2] also tells us that Henry attended the operating theatre of St. Bartholomew's Hospital regularly on Saturdays, a factor in his subsequent career.

After 18 months of study, Frankland took Henry Armstrong into his private laboratory as his personal assistant and it was there that Henry began his first research. Frankland had been appointed as one of three to a Royal Commission tasked with enquiry into the pollution of England's rivers and waterways, so Armstrong's task was to devise ways of estimating the organic impurities in sewage and of sewage matter in drinking water. For this, Armstrong and Frankland established a successful combustion method in vacuum that led to Frankland's analysis of the British water supply. Not surprisingly, after about one year, Armstrong left to study for his Ph.D. degree. Kolbe, in Leipzig, had been mentor to Frankland and it was to him that Frankland sent Henry. There, in October 1867, Henry was introduced to the sulfonic acids and interest in this class of compounds remained throughout his life. In fact, the day that Henry arrived was spent nitrating 4-hydroxyphenylsulfonate (**1**) [2]. He spent the next five semesters studying under Kolbe, the beginning of his lifelong study of aromatic compounds. The master's 'Try it, Try it' had a lasting impact on Henry, who gained his Ph.D. in 1869 for a thesis of which the essence was published in the *Proceedings of the Royal Society* that year [3]. Armstrong returned to London in 1870 taking up residence back in his parents' home, though they had moved to another house on Belmont Hill. There, living next door but one in her parents' home, he met Frances Louise Lavers, who he subsequently married on August 30, 1877.

---

[1] For Sir Edward Frankland, see Chap. 10.

3 Henry Edward Armstrong (1848–1937)

On his return to London, Henry gained a position as assistant to Dr. Matthesson at St. Bartholomew's Hospital, teaching chemistry to the medical students. He held this post for 12 years until Matthesson died. Then, in 1871, he gained another part-time appointment as Professor of Chemistry at the London Institution where he taught classes from 6 to 8 pm on 'analytical chemistry and methods of original investigation' [1, 2]. The salary was low (£50 pa) but he had a private laboratory though no allowance for it or the necessary supplies! The lab, little more than a coalhole, was close to the lecture theatre and it was there that Henry Armstrong began his extensive studies of the constituents of coal tar. The rather unpleasant odours from the higher boiling fractions had a lasting impact on the pupils and teachers using the lecture room. The income from his two posts was insufficient to maintain a professional scientist, let alone the prospects of house and family. Thus, Henry supplemented his income by writing, examining, abstracting (for the Chemical Society Journal) and as a professional witness in legal chemical technology cases. It was from the exposure to examination and cross-examination protocols and the need for precision in speaking, which, when coupled with Kolbe's incisive style, made Armstrong the unpleasant critic that he described himself to be [1, 2]. The various part-time occupations resulted in Armstrong regarding himself as a 'free-lance and iconoclast' [1].

Armstrong's life work began in 1879, some 16 months after his marriage. At age 31, he was appointed by the City and Guilds (of London Institute for the Advancement of Technical Education) to organise classes in chemistry and physics along with William Edward Aryton, a physicist and electrical engineer. The appointments stemmed from lengthy discussions after the 1851 Exhibition and a definite 1876 decision to explore Technical Education by the City Livery Companies of London. It led to the establishment of the City and Guilds Institute in 1878 and Henry's appointment. The classes by Armstrong and Ayrton in Cowper Street, Finsbury, were, in essence, a trial to see how their subjects could best be taught. The fact is they became so popular that a separate specially adapted building was needed. The Finsbury Technical College was created as the first institution of its kind in the city. However, the focus of the City and Guilds was the establishment of the Central Institution (later the Central Technical College) and the expertise of Armstrong and Ayrton was utilised to the full in realising this. The pair went on a tour of laboratories in Germany during the autumn of 1881, each to study the nature of those for his subject. It was from these times that Armstrong's lifelong interest and impact on the teaching of chemistry is dated.

Following the Armstrong-Ayrton report, changes were made to the Central Institute, and the Prince of Wales opened then the Central Institution for more advanced study (in Exhibition Road, South Kensington) in 1884. Henry Armstrong was appointed as inaugural Professor of Chemistry and, after some persuasion, William Edward Aryton the inaugural Professor of Physics. There were two other professors, namely W. C. Unwin (Engineering) and O. Henrici (Mathematics). Armstrong remained until it became a part of the Imperial College of Science and Technology in 1911. By then, 27 years later, Armstrong was 63 years old and required to retire upon the closure. He was given the title of Emeritus and the use of a private laboratory; all his students had left by 1914.

Armstrong's work falls into three categories, (1) the chemical research that started at the London Institution, (2) his interest in and impact on chemical education, and (3) his other activities almost always related to chemistry. It was from Armstrong's chemical researches that his name was recognised far beyond England. His studies, the most substantial covering the chemistry of naphthalene, started in the mid-1870s and provided some 60 short papers, mostly published in the *Proceedings of the Chemical Society*. At its inception, little was known about naphthalene, its derivatives or its chemistry, and industrialists in the dyestuffs industry were only just beginning to see the potential of its compounds. Conceived when Henry was intimately involved in establishing technical education for students in industry, the very extensive and analytically detailed examinations of sulfonation and other reactions, delineated the nature of the ring and the orientation of substitution, results that were of immense value to the dyestuffs industries [1]. The difficulties experienced in separating mixtures of sulfonic acids led Armstrong and Wynne to use chlorosulfonic (chlorosulfuric) acid as their sulfonating agent for naphthalene; they obtained a single product, naphthalene-1-sulfonic acid (**2**; Fig. 3.2a) When **1** was treated with slightly in excess of two molar equivalents of chlorosulfonic acid, diacid **3** was obtained and identified as the 1,5-derivative.

In order to complete this series of studies accurately, Armstrong chose to set a series of reference compounds and, in so doing, he prepared and characterised all

Fig. 3.2 **a** Compounds 1–4; **b** the 10 dichloronaphthalenes with their 1888 mp data (see Ref. [5]); **c** the centric formulae proposed by Armstrong

ten theoretically possible dichloronaphthalenes (Fig. 3.2), eliminating two that had been proposed incorrectly (Fig. 3.2b [4]). The study was experimentally demanding, requiring high experimental accuracy. Many of the compounds were confirmed by synthesising them from different starting compounds of known configuration, a masterpiece of synthetic and analytical accuracy from the late 19th century. The studies led to explanations for the speeds (rates) of certain reactions that included the industrially important diazotisations. By the end of 1895, Armstrong and Wynne had prepared and characterised all 14 of the trichloronaphthalenes, four of which melt over the narrow 90–92 °C range. Each was synthesised in a quantity of between 35 and 50 g. All of his work on the naphthalenes was geared to establishing the factors that govern substitution in the nucleus.

Armstrong was just as interested in the benzene series of compounds and, although much of the nature of substitution had been established by the time he was able to contribute, his study of the sulfonation of aniline and its derivatives had impact as it established that *meta*-directing groups have no *ortho/para*-directing influence and actually inhibit it. By way of example, Armstrong and his students showed that when amines are converted into salts their o/p effect is negated and that, in strong fuming sulfuric acid, metanilic acid (3-aminobenzenesulfonic acid) is obtained. With Miss Evans, he showed that N,N-dimethylaniline gives the p-sulfonic acid with chlorosulfonic acid, whereas the m-acid is formed only with increasing difficulty, and then only when fuming sulfuric acid is the sulfonating agent. It was Armstrong's concepts of the benzene nucleus and his extension to the benzo-fused derivatives that gained attention and it has satisfaction associated with it. Armstrong found that the Kekulé formula for benzene did not provide him with an adequate mental picture of the molecule free from the ambiguity of formal p–bond character. In 1887, he suggested that the molecule was better represented by a 'centric' representation as had been proposed [5] by Lothar Meyer in 1872. Here Armstrong suggested that the fourth valency of each carbon atom was directed towards the centre of the molecule resulting in increased centric density. The six central affinities were assumed to neutralise each other without implying any cross-linking within the ring. Remember, the electron was neither named nor discovered until the 1890s and Armstrong's 'affinity' was, in essence, the electron. The centric form was independently proposed by Baeyer in 1892 as the Armstrong-Baeyer centric formula and it was extended satisfyingly by Armstrong to the annulated derivatives, naphthalene, anthracene and 9,10-anthraquinone as shown in Fig. 3.2c. Henry essentially invented the concept of delocalisation when he said that any one atom has an influence on other atoms not contiguous to it in the ring [6]. These late 1880 ideas were so close to the modern theory of chemistry that it is surprising he never drew the appropriate conclusions once the electron was known [6].

Another important aspect of Armstrong's legacy is his inclusion of crystallography in teaching. He instilled into his students the need to think not just in one but in three-dimensions. He took crystal properties to the chemist and showed that it did not belong solely in the realm of the mineralogist. His own classes at the Central Technical College were the first to impart the ideas and subsequently he arranged for courses on crystallography to be taught from 1886. His publications invariably include a detailed description of the crystals of each new compound reported.

Armstrong made significant contributions to the dyestuffs industry as has already been alluded to. He also instigated the quinone theory of colour, which became a guiding principle to the industry [7]. Although his experimental work in this area was minimal, he proposed that "the unsaturated hydrocarbons are not only more reactive than the paraffins but the beginnings of colour are manifest in them if examination be made in the regions above and below the visible spectrum". Then that "the quinonoid origin of visible colour appeared so general that if a coloured compound was not quinonoid its formula was suspect". More importantly, he claimed that "ultimately colour would be traced to that peculiar condition represented conventionally by a double bond, the atoms being regarded as altogether subordinate". All this before the electron was known or conjugation recognised!

Other inspirational insights from Armstrong come from his 19th century writings that predate chemical discovery of the facts. Thus, he believed that in water HCl interacted with the solvent in the same way as did ammonia, that all chemical change was electrical in nature (a forethought to reaction mechanism?), that mixtures of pure hydrogen and pure oxygen would not be explosive, and that atmospheric corrosion (rusting) would be impossible even in the presence of air and water if an electrolyte were absent.

The educational impact of Henry Edward Armstrong was especially significant to chemistry in the last 20 years of the 19th century and laid the foundation for teaching in the first half of the 20th century. Henry concluded early in his career that the deficiencies in the knowledge of the average boy coming into advanced classes were a result of bad elementary education. He was convinced that higher education in technical subjects could only succeed from a sound educational basis. Thus, in 1884 he proposed teaching the general science of daily life at school by having students perform simple practical experiments. This formulated his heuristic approach to teaching—learn from doing, not simply from a static dogmatic collection of facts—a revolutionary concept at the time. Henry led his students through experiment to self-discovery. He was quoted as saying: *"If the Almighty were in the one hand to offer me Truth and in the other the Search after Truth, I would humbly but firmly choose the Search after Truth"*. He was among the first to base instruction and writing in chemistry upon Mendeleev's periodic table and, early on, he emphasised that molecules must have spatial configurations that determine crystal structures.

From 1887 Armstrong served on two important education committees that of the British Association charged with reporting on teaching methods, the other on the teaching of chemistry. The reports were influential [8] in bringing about the changes that Armstrong saw as vital for science (and chemistry) but it took many years before the recommendations had any real effect. Nevertheless, it was Armstrong who had the foresight to see what was needed and it is he to whom science education is much indebted. Henry fought the conservative spirit that pervaded much of society all of his life. That his views were still not fully accepted, even after his death, is nicely

3 Henry Edward Armstrong (1848–1937)

illustrated [1] by a 1940 letter from Plymouth school teacher A. G. Lowndes in Nature. Lowndes states [9]:

> As pointed out by the Editors of NATURE a large majority of the boys complete their 'formal' education at the School Certificate stage and for this 'formal' education practical work is unnecessary. Had I my own way,..... I would make every boy in a public school learn Latin until he were either fifteen or had passed the School Certificate. Two good science lectures weekly with demonstrations are all that are required.

At the Central Institution in South Kensington, Henry lectured only to the first-year students espousing the view that the ground-work was the most important, a feature common in Commonwealth Universities until the latter part of the 20th century. His lectures covered much ground and were filled with demonstrations that almost always worked. For the senior levels, he led discussions rather than held formal lectures (the modern tutorial!). Armstrong was a Governor of St. Dunstan's College in Catford, neighbouring on to Lewisham, and Christ's Hospital, a co-educational school established in 1552 for the orphan children of poor Londoners in the old Grey Friars buildings on Newgate Street in the City. To this latter school he gave great service, especially after its removal to West Horsham in Sussex in 1902 when he designed and equipped its laboratories. It was the best school for science teaching in England. Henry also set up workshops for manual training and revolutionised art training. Armstrong was deputy chairman of Christ's education committee for 14 years from 1926 and its Chairman until 1937. There can be no doubt that significant changes did take place in the 50+ years after Armstrong began his educational campaign. That they happened at all are in very great measure due to him.

Henry Armstrong also had much interest in agriculture that likely started with his upbringing in rural Lewisham. This impacted on his chemistry and he made major contributions to terpene and camphor research that date from 1878 and involved the isolation of the D- and L-forms of sobrerol (**4**, Fig. 3.2a, which he named after Ascanio Sobrero, its 1851 discoverer). He became associated with the Rothamstead Experimental Station in Harpenden, Hertfordshire. He served as the Chemical Society representative of the Lawes Agricultural Trust Committee from its establishment in 1889 until shortly before his death in 1937. His efforts persuaded the Indian Government to set up a research institute for the cultivation of indigo in 1915. His strong support of the British Dyestuffs industry extended to his personal proffering of its wares. He wore brightly dyed waistcoats at formal evening functions. One such piece of clothing, a Caledon Jade Green, has been described as the most becoming, but another of indigo-blue with thioindigo-red (Fig. 3.1) facings was the most striking. Both are Vat Dyes that had to be reduced to make them inert and water-soluble prior to the dyeing process, then oxidised to regenerate the original colour.

Henry Armstrong was a strong supporter of the Royal Society of Arts [10]. The first meeting that he presided at was in 1883 and he occupied the chair seven times. He was awarded the Society's silver medal for a lecture on the indigo situation in India and its Albert Medal for his discoveries in chemistry and his services to education.

Much of what Henry Edward Armstrong foresaw has come true and many of his beliefs of what should be studied formed the basis of much late 20th century research.

Several remain current areas of intense activity and include dyes and dying (now nanoparticle involvement), agricultural chemistry, diet and nutrition, and collaborations and joint interdisciplinary publication. In many respects, Henry Armstrong was a man before his time. His educational writings are listed for completion [11–15].

Armstrong was awarded a Hon. LLD from St. Andrews, Hon. D.Sc. degrees from Melbourne and Madrid, the Davy medal of the Royal Society, the Messel medal of the Society of Chemical Industry, the Albert medal (see above), and the Horace Brown medal of the Society of Brewing. He served the Chemical Society for many years ruling it with 'an iron rod' [1] as Secretary from 1875 to 1893 and serving as President from then until 1895. Following this tenure, he took the role of Vice-President until his death save for two short breaks. He was, in essence, a member of the Council almost continuously from 1873 until 1937.

Upon marriage to Frances Lavers in 1877, Armstrong purchased his own home in Lewisham—one that his father had previously owned! The couple had seven children, four boys of whom industrial chemist Edward Frankland Armstrong (1878–1945) was the eldest, and three daughters. In 1882, the family moved to Granville Park in the same town and that was their permanent home. Their youngest daughter Nora never married but looked after her parents until their deaths, Frances in 1935 and Henry some two years later on July 13 in 1937.

References

1. Rodd EH (1940) Obituary—Henry Edward Armstrong. J Chem Soc 1418–1439
2. Keeble FW (1941) Henry Edward Armstrong. 1848–1937. Obit Not Fellows R Soc 3(9):229–226
3. Armstrong HE (1870) Contributions to the history of the acids of the sulphur series—I. On the action of sulphuric anhydride on several chlorine and sulphur compounds. Proc Royal Soc 18:502–513
4. Armstrong HE, Wynne WP (1890) The ten isomeric dichloronaphthalenes and the sulphonic acids and trichloronaphthalenes derived therefrom. Proc Chem Soc 6:77–88
5. Meyer L (1872) Die modernen theorien der Chemie und ihre Beteutung für die chemische Mechanik, 2nd edn. Maruschke und Berendt, Breslau
6. Rzepa H. Chemistry with a twist. www.ch.imperial.ac.uk/rzepa/blog/?paged=24. Accessed 21 Nov 2014
7. Armstrong HE (1888) The origin of colour and the constitution of Colouring Matters. Proc Chem Soc 4(49):27–34
8. Brock WH (ed) (1973) H. E. Armstrong and the teaching of Science 1880–1930. Cambridge University Press, Cambridge
9. Lowndes AG (1940) Practical science in schools. Nature 146(3691):133
10. Evans EV, Woolcock JW (1937) The Late Professor H. E. Armstrong, F. R. S. J Royal Soc Arts 85(4421):903–904
11. Armstrong HE (1884) On the teaching of natural science as a part of the ordinary school course, and on the method of teaching chemistry in the introductory course in science classes, schools, and colleges: address for the proceedings of the international conference on education, London
12. Armstrong HE (1874) Introduction to the study of organic chemistry: the chemistry of carbon and its compounds. Longmans, Green, and Co., London
13. Armstrong HE (1903) The teaching of scientific method and other papers on education. Macmillan and Company, New York

References

14. Armstrong HE (1925) The first epistle of henry the chemist to the Uesanians. J Chem Educ 2:731–736
15. Armstrong HE (1927) Art and principles of chemistry. Ernest Benn Ltd., London

Chapter 4
Leo Henricus Arthur Baekeland (1863–1944)

Leo Henricus (Hendrik) Arthur Baekeland (Fig. 4.1) was born in Ghent in Belgium on November 14, 1863. He was the son of Charles, an illiterate cobbler, and Rosalie (née Merchie), a domestic servant. At the age of five, he entered primary school and then, at age 13, his father apprenticed him to another shoemaker. However, his mother, whose work took her to the well-to-do households of Ghent, insisted that her son attend a secondary school to gain a better education with improved prospects; Leo entered the Atheneum. When old enough, he enrolled at the Municipal Technical School, where he took evening classes in chemistry, physics, mechanics and economics, winning a medal in each of the four subjects. It was there that his lifelong commitment to chemistry started. In fact, his promise was such that the City of Ghent awarded him a scholarship to attend university. He accepted and entered the University of Ghent in 1880 at age 17, taking natural sciences and specialising in chemistry under Professor Theodore Swarts, who had succeeded Friedrich August Kekulé in 1867. Although the youngest member of his class, Leo was the most gifted. He graduated with a B.Sc. after two years, and a Ph.D. involving electrochemical studies maxima cum laude only two years later, ahead of his 21st birthday. During this time, Leo supported himself with the scholarship from Ghent, by serving as a lecture assistant, and by teaching to reduce the contribution from his parents [1–3].

Young Baekeland had a boyhood interest in photography. This led him to dissolve the silver chain of his pocket watch in nitric acid because he needed silver nitrate but could not afford to buy it. The resulting solution also contained copper nitrate and, as one of his first chemistry projects, he worked out a separation of the copper from the silver salts [2]. Following his doctoral graduation, Leo taught chemistry and physics at the Government Higher Normal School of Science in Bruges. There, he continued his interest in photography having been encouraged while a student by photographer-chemist Désiré van Monkhoven. During this period he developed a technology with economic potential—the dry bromide photographic plate. His plates

---

A previous version of this chapter was published as Halton B (2016) Leo Hendrick Baekeland (1863–1944). Chem N Z 80:148–153.

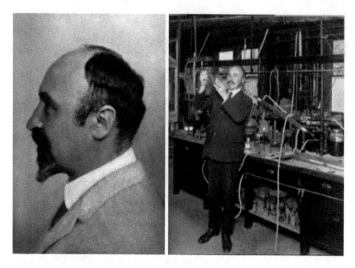

Fig. 4.1 Left: Dr. Leo Henricus Arthur Baekeland, an American scientist born in Belgium (Preface of The World's Work. Garden City, New York: Doubleday, Page & Co. via Wikipedia from 1916 image enhanced by Reindert Groot); Right: Leo Baekeland in his laboratory (image provided by Reindert Groot at www.amsterdambakelitecollection.com)

carried a light-sensitive material and a coating of dry developer and a protective layer to prevent oxidation. Developed in water immediately after exposure, negated the hassles with plates that had to be coated wet just before exposure [4]. He returned to Ghent in 1887 as Assistant Professor where he combined a promising academic career with his industrial activities; he gained a patent for the dry plate. Together with his colleague Jules (and his wife Valérie) Gleesener, he established *Dr. Baekeland et Compagnie* to manufacture the dry plates. Even at the age of 24, it was clear that Baekeland brought the right technology to the company, while Mde Gleesener made the infrastructure and capital available. Sadly for Baekeland, the company failed due to the difficulty of combining academic work with industrial activity and the fact that the era of plate technology, wet or dry was coming to its end. Baekeland and Swarts quarrelled frequently because the senior wanted his former brilliant student and now a junior colleague to become a famed academic.

Baekeland returned to Ghent in 1887 to be closer and better able to court Celine Swarts, the professor's daughter who he had met while a student. That year the four universities of Belgium (Ghent, Louvain, Liege and Université libre de Bruxelles) held a competition among its graduates of the previous three years and awarded Baekeland the first prize. This provided him with a title, a gold medal and a travelling scholarship. Therefore, following his promotion to Associate Professor, he visited Berlin, London, Oxford, and Edinburgh Universities before returning to Ghent where he married Celine on August 8, 1889. Two days later the newlyweds travelled to New York and honeymooned. There he met Richard Anthony of the photographic supply house E. and H.T. Anthony & Co., makers of dry plates and bromide paper. Richard introduced Leo to Columbia University's Chemistry Professor, Charles F. Chandler, a chemical consultant to the company, an enthusiastic amateur photographer, and a force in late 19th century American chemistry. Chandler saw in Baekeland a considerable talent and persuaded him to remain in the US to solve industrial problems. Given this advice, Leo resigned from his position in Ghent, and joined the Anthony Company as their chemist. There, he worked on film emulsions and printing papers. However, he stayed with them for only two years, choosing to become an independent consultant and research chemist in 1891 [4]. This was after the birth of his first daughter Jennifer in 1890. She died of pneumonia in 1895.

The move to consulting was, perhaps, Baekeland's biggest failing as he tried to develop too many processes too quickly; he became ill and his finances dropped to their lowest. During his recovery, he decided to concentrate on one single well-chosen project to give him the best chance for the quickest possible results [2]. With his exceptional grasp of the principles of basic science, he then proceeded to produce results that revolutionised several major branches and had an impact on the day-to-day lives of millions. They made him a wealthy celebrity [5]. The first success involved a return to his photographic interests with the development of the Velox photographic paper [4]. Baekeland's objective was to produce a better photographic paper than then available and one capable of being developed in artificial light. His *modus operandi* was to think of a problem and then devise an experiment to prove it wrong. If that experiment did not negate his idea, he would then experiment meticulously and accurately to gain a new and meaningful solution—a 'look for a problem, get a better solution and commercialise it' approach. After two years of intensive experimentation that included many failures, he perfected a process to his satisfaction and produced a paper using chloride rather than bromide; he named it Velox. He had generated more than 50 variants of his chloride paper. Each was carefully assessed for coatability, sensitivity, tonal quality, shelf life, and image lifetime. Velox gave better light sensitivity with a variability in tonal quality that gave softer prints. However, at that time, the US was suffering a recession and there were no investors or buyers for his new superior product despite the growing popularity of the reloadable camera and the increased need for print developing.

Fortunately, Leo Baekeland found a commercial sponsor for his paper, one Leonardi Jacobi, a scrap metal dealer from San Francisco, and they established the Nepera Chemical Company in Nepera Park, Yonkers, New York. However, when set to produce quantities of photographic paper for market, his factory was beset with summertime humidity problems that caused the emulsions to become sticky, pre-venting application for weeks at a time. Baekeland devised the solution (air conditioning for the plant). He had the external air passed over ice to condense the moisture sufficiently to allow plant operation. The company began producing paper in 1893. Within a few years, Velox had taken a large part of the market; the prints were durable and had excellent tone with rich blue-black colours. Furthermore, Baekeland offered Velox in a range of grades for photo finishers to make the best prints from thin or heavy negatives. Most important was that Velox was easy to use and gave reliable results as it was designed for use in artificial light—it gave rise to the *Back-of-the-Pharmacy* processing and was the first commercially successful photographic

Fig. 4.2 Left: Snug Rock (ca. 1900), Baekeland's home in Yonkers New York (from Bob Piwinski's Victorian). *Source* See Ref. [3]; Right: Baekeland's Laboratory at Snug Rock, Yonkers, New York (file provided to Wikimedia Commons by the Science History Institute)

paper. Rather than expand the plant and hope to retain his market share, Leo decided to sell and in 1899 he, Jacobi, and Albert Hahn (a further associate) sold Nepera to George Eastman of the Eastman Kodak Co. for $750,000. Baekeland earned approx. $215,000 ($US6.5 M today) from the transaction and with a portion of the money, he purchased Snug Rock (Fig. 4.2), a handsome estate in the then Harmony Park section of Yonkers overlooking the Hudson River. He also bought a car to satisfy his motoring hobby much to the chagrin of his horse-owning neighbours who thought him a gasoline devil.

Although Baekeland was financially set for life with little need to work, he set up his own well-equipped laboratory in what had been a barn on the property and continued his researches there. However, part of his contract with Eastman was not to perform photographic studies for at least 20 years; he had to find a new area of research. He returned to electrochemistry as it had begun to revolutionise chemical industry with, for example, the separation of aluminium from bauxite. He spent the winter of 1900 at the Technical Institute at Charlottenburg (now The Technical University of Berlin) improving his knowledge of the subject. On returning to the US and Yonkers, he equipped his laboratory for electrochemical study. Shortly, Elon Huntington Hooker, who had set up the Hooker Chemical Company in 1903 as a Development and Funding Company, asked Leo to assess the recently invented Townsend electrolytic cell. This was with a view to possible industrial electrolysis of brine for chlorine and caustic soda production [2]. The cell, invented by Clinton Townsend and Elmer Sperry (subsequent founder of Sperry Electric) comprised of a tank in which electrolysis generated sodium hydroxide (caustic soda), chlorine, and hydrogen. Townsend had placed a diaphragm between the anode and cathode, had a perforated kerosene-filled jacket around the cell, and devised a means of keeping the chlorine, hydrogen and caustic soda separate; its operation needed a plentiful supply of power. Baekeland's work resulted in improvements to the diaphragm, a reduction in corrosion, and higher overall electrical efficiency. Together Leo and Hooker built a pilot plant in Brooklyn at a cost of $300,000. The electrolytic process was honed and made ready for commercialization subsequently saving them far more than the pilot plant cost. Full production began at Niagara Falls in 1905 and utilised the low cost electricity from the Niagara Falls power project, water from the Niagara River and, importantly, the abundance of salt from the nearby mines. It was one of the world's largest electrochemical plants. The Hooker Chemical Co. subsequently manufactured a range of chemicals from the electrolytic products. Undoubtedly, Leo Baekeland added to his wealth and it provided for one of his more famous quotations: *Commit your blunders on a small scale and make your profits on a large scale.*

Following the electrolysis of brine, Leo Baekeland turned his attention to organic chemistry and the condensation of formaldehyde with phenols from recognizing the potential of, already known but not commercial. The initial impetus came from the idea of replacing natural shellac with a manmade alternative. Shellac is a resin secreted by the female lac bug on trees in the forests of India and Thailand and used as a natural primer, stain, and a high-gloss varnish also used in electrical applications) with a manmade alternative. The work led to Baekeland's greatest success—the discovery of Bakelite. Von Bayer had described experiments with phenol and formaldehyde some 30 years earlier and a number of other recognised chemists had tried to find useful products from the reactions. Amongst others, Austrian chemist Adolf Luft gained a patent (1902) for a process to produce plastic materials from phenol-formaldehyde resins especially suitable as a celluloid substitute. However, the amber-like material that he obtained from adding camphor was too brittle for technical use. Despite these studies, no one had produced a viable product [6]. The scene was set for the systematic examination that Baekeland provided.

As was his norm, Leo began his studies by repeating all of the earlier work [1, 2, 4]. Together with his assistant Nathaniel Thurlow (he rarely had more than two) the same useless resinous masses were obtained. Then began systematic variation of each individual component of the reaction between phenol and formaldehyde. Thus, the aldehyde, the substitution on the phenol, the stoichiometry, the temperature, and the pressure were varied. Alkali, ammonia and other bases replaced the acid, and the effect of each individual change on the reaction noted. Only after some five years of intensive effort did success come. Baekeland had found that pressure, in particular, impacted on the process and he was able to control the reaction at its various stages, separate it into different steps, and slow it down by adding ammonia. Ultimately, he was able to stop the reaction at almost any stage [2]. We now know that phenol (**1**) and formaldehyde (**2**) react in acid to give condensation product 4-hydroxymethyl-phenol (**3**) as shown in Fig. 4.3. This, on protonation and further reaction with phenol gives 4,4'-methylenebisphenol (**4**), in analogy to the ubiquitous bisphenol A. As phenol is an *ortho/para* directing group reaction at the less reactive *ortho*-position is also possible. As the reaction sequence proceeds, this *ortho*-coupling comes into play generating a cross-linked polymer product. Baekeland's product was just this with the essential core, 5, illustrated in Fig. 4.3. It was the world's first completely synthetic plastic and named Bakelite (polyoxybenzyl-methyleneglycol) by its inventor. The first patent was filed on July 13, 1907 and has become known as the '699 patent [7], the first of some 400 that Bakeland was granted.

Despite the laboratory success, Baekeland spent two further years having the product tested by various potential users so that the processes could be refined for commercial production. It was at this time the sealed autoclave used in the Snug Rock

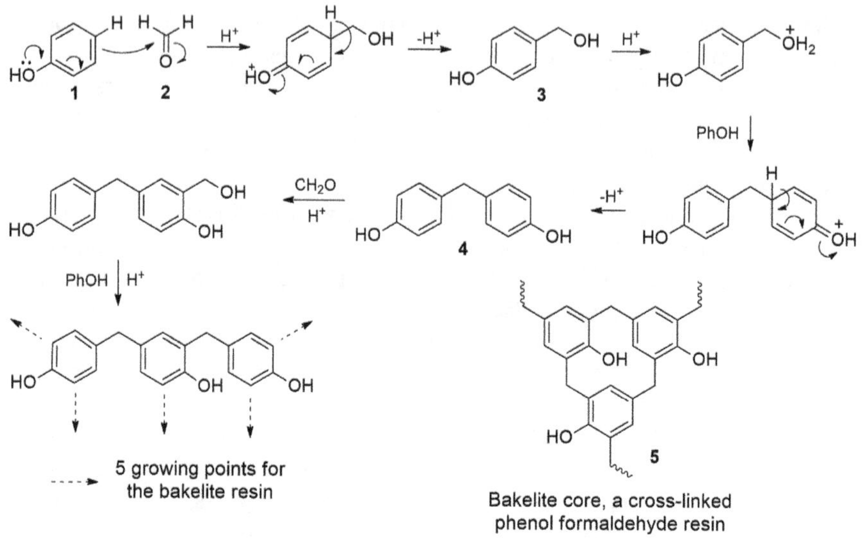

Fig. 4.3 The phenol-formaldehyde reaction showing only *para-para* coupling; *ortho-para* and *ortho-ortho* are also possible

lab became known as *Old Faithful* (Fig. 4.4); years later in 1983, it was rehoused in the Smithsonian Museum. Only after these trials was commercialisation set, with his results publically disclosed at the February 8, 1909 meeting of the New York section of the American Chemical Society. In his presentation, Leo said: *Previous reactions had resulted in slow processes and brittle products.... But by use of small amounts of bases I have succeeded in preparing a solid initial condensation product, the properties of which simplify enormously all molding operations.* Because users had difficulty in completing the final steps of production, Leo started semi-commercial production in his laboratory and, in 1910, when daily output had reached 180 litres (mainly for electrical insulators) he formed *General Bakelite* (1911) and began operations in Perth Amboy, New Jersey, to manufacture and market his new material. By 1930, this was *The Bakelite Corporation* with a 128-acre plant at Bound Brook, New Jersey. In 1979 at age 73 years of age and prompted to retire by his son, Baekeland sold the company in 1939 to Union Carbide.

Bakelite was not the first plastic, as celluloid—ultimately derived from cotton and other vegetable matter—had been around for some time. What was so special about Baekeland's phenol-formaldehyde resin was that it softened on heating allowing molding and it was capable of dissolution. However, on reheating to higher temperature it set into a permanently hard and insoluble heat-resistant material unaffected by many chemicals. It was tough, stable and fire resistant, and was less expensive and markedly more versatile than any cellulosic material of the day. Most importantly, it was the first totally synthetic plastic and it revolutionised the manufacturing of everything from buttons to car parts. What made Bakelite so popular was its

Fig. 4.4 Old Faithful—Baekeland's first semi-commercial unit installed in his Yonkers laboratory and later shipped to Perth Amboy for making small batches. From 1983 it has occupied a place of honor in the National Museum of American History (image from the Science History Institute, as part of the Wikipedian in Residence initiative)

multifaceted and extremely useful properties. It was a thermosetting resin moldable and high temperature resistant, and it retained its shape. Its resistance to electricity and chemical action made it a great material for electrical insulation and in appliances. *Time* magazine put Leo Baekeland on its September 22, 1924 cover [8] with the article stating: *From the time that a man brushes his teeth in the morning with a Bakelite-handled brush until the moment when he . . . falls back upon a Bakelite bed, all that he touches, sees, uses will be made of this material of a thousand purposes.* This is not far from the truth as Bakelite encompassed the world for years with its products considered synonymous with high quality and durability—essentially becoming the 1920s and 1930s equivalent of today's Intel Inside. Some of the more notable uses of Bakelite with their date of introduction are:

1912—Hyat Burroughs replaced celluloid by Bakelite in billiard balls.
1914—The Bakelite telephone receiver case was made by Western Electric.
1915—The US Eastman Kodak camera was encased in Bakelite.
1923—The Philips radio company started its own Bakelite production (Philite in The Netherlands) with the first radio speaker in production in 1927.
1926—The first Bakelite chair appeared.

Bakelite was manufactured in several forms to suit varying requirements but the Bakelite resin was fundamental to all. Until the patent expired in 1927, Bakelite was used mostly in industrial applications, such as automobile and electrical insulators. However, once the patent lapsed, it broke into more widespread use. In October 1925, the first issue of *Plastics* magazine gave the colour range available with the variety including Clear Material, for jewellery, smokers' articles, cement, used in sealing electric light bulbs in metal bases; varnishes, for impregnating electric coils; lacquers, for protecting the surface of hardware; enamels, for giving resistive coating to industrial equipment. Laminated Bakelite, was used for silent gears and insulation; and molding material, from which are formed innumerable articles of utility and beauty. The year 1927 was the turning point in the use of Bakelite because the '699 patent expired and real competition with the Bakelite material was then possible.

Bakelite should not be confused with Catalin, another thermosetting phenol-formaldehyde resin popular from the 1930s. It was developed and trademarked in 1927 by the American Catalin Corporation after the Bakelite patents had expired. It is produced in a different two-stage manufacturing process than other types of phenolic resins, contains no fillers and is near colorless, rather than opaque. Unlike other phenolics, it can be produced in bright colours or even marbled. This made Catalin more popular than other types of Bakelite for consumer products. Most Bakelite jewellery for sale is actually Catalin.

Phenol-formaldehyde resins are still in use. With an appropriate catalyst the condensation reactions of Fig. 4.3 can be maximized for *ortho-ortho* or *ortho-para* coupling [the latter using base and dihydroxymethane, $CH_2(OH)_2$]. Such use has given rise to the Novolac and Resol resins of, for example the plastics engineering companies Plenco [9], and the Sumitomo Bakelite Co. Ltd. based in Japan [10]. Even as late as 1995 one of Baekeland's inventions made the news as the heat shield on the 1995 Jupiter probe was phenol-formaldehyde resin-based.

Leo Baekeland became a professor by special appointment at Columbia University in 1916, held numerous honorary degrees and awards, and was a member of all the relevant professional societies. He was the 1924 President of the American Chemical Society. After the success of Bakelite, Leo spent much time in his lab on various academic projects maintaining his active interest in research. In retirement, he spent the winters at his estate in Coconut Grove, Florida but as he aged his mind began to fail. He died of a cerebral hemorrhage in a sanatorium in Beacon, New York on February 23, 1944. He is buried in the cemetery at Sleepy Hollow.

Leo and Celine Baekeland had three children, daughter Jennifer referred to above and who appears in only some of the genealogical files, Nina (b.1896, d.1975) and George Washington (b.1895, d.1966), who never lived up to the high expectations of his father. Much of the Baekeland inherited wealth was subsequently squandered by his descendants. Indeed the history of the Baekeland family after Leo's death has been equated to something of a crime novel as the socialite wife Barbara Daly Baekeland of his grandson (George Middleton "Brooks" Baekeland, son of George W.), was murdered by her 26 year old homosexual son Anthony in 1972 [11]. Barbara's high society life was the subject of the 2007 film Savage Grace based on the book by Robins and Aronson [12].

References

1. National Historic Chemical Landmarks program of the American Chemical Society (1993) The Bakelizer. https://www.acs.org/content/dam/acsorg/education/whatischemistry/land-marks/bakelite/the-bakelizer-commemorative-booklet.pdf. Accessed 9 December 2015

References

2. Kettering CF (1946) Leo Hendrik Baekeland 1863–1944. Biogr Mem (Natl Acad Sci USA) 24:281–302
3. Victorian Source. Dr. Leo Hendrik Baekeland. www.victoriansource.com/id34.html. Accessed 27 Jan 2016
4. Kaufmann CB (2011) Leo H. Baekeland. In: Strom ET, Rasmussen SC (eds) 100+ years of plastics. Leo Baekeland and Beyond. ACS symposium series 1080, American Chemical Society, Washington DC, pp 1–10
5. Thomas AW, Burke SP, Fink CG, Turner WD, Hixson AW (1944) Leo Hendrik Baekeland. Science 100(2585):22–24
6. Weber WE (1995) Of bicycles, bakelites, and bulbs: toward a theory of sociotechnical change. MIT Press, Cambridge
7. Baekeland LH (1909) Method of making insoluble products of phenol and formaldehyde, US Pat. No. 942,699 (filed on July 13, 1907, issued December 7 1909)
8. Cover (1924) Time 4(12), 22 Sept
9. See e.g. Plenco. https://www.plenco.com/phenolic-novolac-resol-resins.htm. Accessed 4 Feb 2016
10. Sumitomo Bakelite Co. Ltd. http://www.sumibe.co.jp/english/product/index.html. Accessed 5 Feb 2016
11. See the book review: Leafe D (2008) How a society beauty was finally murdered by the gay son she had seduced. http://www.dailymail.co.uk/femail/article-1030330/How-society-beauty-finally-murdered-gay-son-tried-cure-homosexuality.html. Accessed 17 Jan 2019
12. Robins N, Aronson SML (2007) Savage Grace, The true story of fatal relations in a rich and famous American family. Simon and Schuster, New York

Chapter 5
Alexander Porfirevich Borodin (1834–1887)

The name Borodin has little or no meaning to many, but to others it conjures up the sounds of the Polovtsian Dances from the opera Prince Igor. The 1953 musical Kismet popularised the themes most notably with "Stranger in Paradise" (credited to Wright and Forrest) that came from the "Gliding Dance of the Maidens" and for which Wright, Forrest and Borodin (posthumously) gained a 1954 Tony Award. Alexander Borodin (Fig. 5.1) was, and remains, a highly respected composer. He was one of the Russian group called "The Mighty Handful" (Fig. 5.2)—"Moguchaya kuchka" in Russian—a group of five Russian composers [Mily Balakirev (leader), César Cui, Modest Mussorgsky, Nikolai Rimsky-Korsakov and Alexander Borodin], who in 1862 banded together in St. Petersburg in an attempt to create a truly national school of Russian music. They were all young self-trained amateurs at that time; Balakirev was 25 years of age, Cui 27, Mussorgsky 23, Borodin the eldest at 28, and Rimsky-Korsakov just 18 years old. Their objective was to free Russian music of the stifling influence of Italian opera, German lieder, and other western European forms. In a sense, they were a branch of the Romantic Nationalist movement.

Despite becoming a recognised and highly respected musician, Borodin simply claimed to be a "Sunday musician" with Monday to Saturday occupied with his profession. His most famous quote is "Respectable people do not write music or make love as a career," but he is also credited with [1]:

> As a composer seeking to remain anonymous, I am shy of confessing my musical activity… For others it is their chief business, the occupation and aim of life. For me it is a relaxation, a pastime which distracts me from my principal business, my professorship. I love my profession and my science. I love the Academy and my pupils, male and female, because to direct the work of young people, one must be close to them.

Alexander Borodin was a chemist!

Alexander Porfirevich Borodin was born on February 12, 1834, in St. Petersburg, the illegitimate son of Prince Luka Gedianov, a 62-year old Imeritian and his 25-year old maid Avdotya (Narva) Konstantinovna Antonova. The baby was born in Gedianov's house and continued to live there with his natural father and mother until 1839, when

© Springer Nature Switzerland AG 2020
B. Halton, *Some Forgotten Chemists*, Perspectives on the History of Chemistry,
https://doi.org/10.1007/978-3-030-16403-4_5

Fig. 5.1 Alexander (from blakedynasty.typepad.com from Wikipedia) and Ekaterina Protopopova (Borodina) Borodin (from The One-2-Five E-tutor from Stallwood Music Composers—Romantic)

Fig. 5.2 The Mighty Handful. Left Mily Balakirev; right clockwise: César Cui, Modest Mussorgsky, Alexander Borodin, and Nikolai Rimsky-Korsakov (from Wikipedia in public domain)

5 Alexander Porfirevich Borodin (1834–1887)

Luka arranged a marriage for Narva to an elderly retired army physician to gain her an inheritance. Avdotya was a woman of the middle classes whom Luka loved but was unable to marry given the class structure of Russia at that time. As was common then, the boy was given the name of one of Gedianov's servants, Porfirevich Borodin, and registered as his son. Thus, Alexander was raised in a privileged household where he always addressed his mother as Auntie or Aunt Mimi. She had two other illegitimate sons with Prince Luka but it seems that they were not given the same treatment [1].

Alexander's father was in receipt of a liberal government pension following the 1810 annexation of Imeritia (Georgia) by Russia and was able to live a life of luxury. He devoted monies for his son's education, which was organised by his mother. Alexander remained his father's serf until 1840 when, sometime before his death Luka Gedianov released him. Alexander was a weak and unhealthy child, a constitution inherited from his parents that remained with him throughout his life. Largely because of this he was home tutored until he was 13 years old. By then he was fluent in German, French and English, and subsequently mastered Italian to the extent that he could write scientific papers in the language. He had an impressive memory and excelled in all academic subjects. As a boy he studied the flute, cello and piano and wrote his first composition when he was nine, the polka Helene that was dedicated to a woman named Elena. He was so talented, he could reproduce on the piano what he had heard a military band play. In addition, he performed creditably on the oboe, the clarinet and several brass instruments. At 13 years of age, he composed a concerto for flute and piano and a trio for two violins and cello. The oriental heritage from his father seems to account for his looks and the themes that pervade so much of his subsequent music. However, it was not so much music as chemistry that was his passion and Alexander fitted up a small laboratory in his room and performed experiments making fireworks and performing chemical magic for his friends. One of these, Mikhail (Misha) Shchiglev (who became a noted music teacher), has said [2]: *Almost the entire apartment was filled with jars, retorts and all sorts of chemicals. Tubes with crystalline solutions were on the windows everywhere. The whole house smelled of his chemical preparations and his visiting teachers were afraid of a fire.* Prince Luka did not think a musical career appropriate for a person of royal lineage and discouraged Alexander from it, pressuring him to a medical career that was supported by the stepfather. Borodin became attracted to organic chemistry.

The young Borodin was sent to school to prepare for his professional career where Misha was also a pupil. Between classes the two boys would go to a piano and play the symphonies of Hayden and Beethoven in duet arrangements [3]. In 1850, Alexander entered the Medico-Surgical Academy of the University of St. Petersburg making friends with many of the German students. Although his coursework was full-time, Borodin still attended concerts and practiced his flute and cello, even spending all night with like-minded friends [4]. In his first year of study, he took classes in botany, zoology, anatomy and crystallography while in his second year he narrowly escaped death from an infection acquired during an autopsy. Throughout this time, his main efforts were in chemistry where the noted Nikolay Nikolaevich Zinin was professor. It was Zinin who discovered the reduction of nitroaromatics to anilines with sodium sulfide (the Zinin reaction) and is the reputed grandfather of

Russian organic chemistry. In his third year, Borodin asked if he could study under his direction and was allowed to do so. However, on one occasion, he spent almost 24 h playing uninterrupted concert music and for this he was severely reprimanded for "trying to hunt two hares at the same time" [3].

In 1856 Borodin completed his courses at the Academy, by which time he had gained the respect of Zinin and the other academics, and was highly praised. He was immediately appointed a junior officer in the Preobrazhensky Regiment, one of the oldest and elite regiments of the Russian army, as assistant in general pathology and therapy at the Second Military-Land Forces Hospital [3, 5]. Apparently, the sight of blood was not to Borodin's taste but he pursued his studies for the completion of his medical degree. In August 1857, he travelled to Brussels with a senior oculist to attend an international ophthalmic congress and then, in March 1858 presented a paper "On the action of ethyl iodide on hydrobenzamide and amarine" to the Russian Academy of Sciences [5], which was published later that year [3] giving him his first paper [6]. This was from work carried out at the military hospital [7]. Two months later he was awarded his medical doctorate by the academy for his thesis "On the analogy of arsenic acid and phosphoric acid in chemical and toxicological behaviour" [5]. Apparently, his was the first thesis at the academy to be written and defended in the Russian language rather than Latin [8].

After a short time, Borodin returned to Zinin's laboratory as his assistant and not too long afterwards was appointed to a position in the academy. Then in 1859, Zinin arranged for him and other young Russian chemists, including Mendeleev, to gain experience in Europe, and this at the government's expense. The idea was to ready the young men for academic careers on their return home. For Borodin, study was to be in Heidelberg with Bunsen but he found the conditions in his laboratory unsuitable and so he worked under Erlenmeyer instead [7]. Here his studies involved the benzidenes, which became important materials in the dye industry, and he published five papers from his 12-hour workdays (starting at 5 am!) [9] in Erlenmeyer's journal, *Zeitcshrift für Chemie und Pharmacie* [10–14]. It is interesting to note that in work published in Russia and France at that time, 6 was used as the molecular weight of carbon whereas the value of 12 was used in Erlenmeyer's laboratory and it appeared in the Heidelberg papers.

The first piece of work published by Borodin [6] defined above (Fig. 5.3a) was his attempt to settle the formulae of hydrobenzamide **1** and amarine **2**, and especially the number of N-H units in each compound. Hydrobenzamide, first synthesised in France in 1836, was known to give amarine on heating to 130 °C. By allowing each compound to separately react with ethyl iodide, we now recognise that the imine nitrogen atoms are quaternised and the H of the >NH group replaced by an ethyl giving rise to the corresponding salts **3** and **4** (Fig. 5.3a). Although Borodin did not reach these conclusions [9], he returned to the compounds later, presenting his final results at a meeting in Kazan in 1873 when he gave the correct number of NH moieties as zero for **1** and one for **2**.[1] This latter work was published by Borodin

---

[1] See Ref. [9], p. 123 and citation 12 therein.

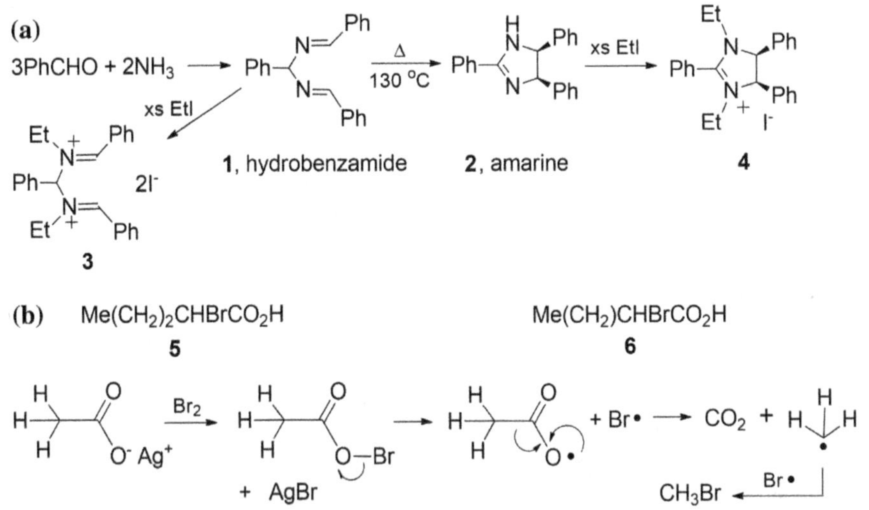

Fig. 5.3 a Benzaldehyde-ammonia reaction and b silver acetate to bromomethane sequence

that year and reported in *Berichte der Deutschen Chemischen Gesellschaft* (1873) by von Richter and from there to the *Journal of the Chemical Society* in 1874.

In his next published work, Borodin reported the syntheses of bromo-butyric **5**, and -valeric **6** acids using the 1836 method of Péligot [15], which involved treating the silver salt of the acid with bromine vapour. He found the compounds decomposed on distillation in contrast with the stability of bromoacetic acid, which he could not prepare in the same way. Treatment of silver acetate with bromine vapour led only to silver bromide and other gaseous products from which carbon dioxide was isolated and the other was thought to be either methyl (or ethyl) bromide (Fig. 4.2b) [9]. This study was reported on a visit to Paris in late 1860, where Adolphe Wurtz was carrying out his studies at the École de Medicine. It was published in French the following year and then in German in *Liebig's Annalen*[2] [16, 17]. The formation of an alkyl bromide from the silver salt of the derived acid has been routinely described as the Hunsdiecker reaction [18] reported only in 1942. Borodin's 1860 contribution more than 80 years earlier, received scant recognition in the chemical literature [9] until recently; the account on *Google* notes that the reaction now also is called the Borodin reaction, the name commonly used in Russia.

Soon after arriving in Heidelberg, Borodin met Dmitri Mendeleev who also was to work with Bunsen. Mendeleev did work with Bunsen but set up his own private laboratory rather than use the facilities available in the university. The two young men became lifelong friends and took holidays together while in Germany. The year after arriving in Heidelberg the pair went on a Southern German holiday with Zinin prior to attending the famous 1861 International Chemical Congress at Karlsruhe

---

[2] See Ref. [9], p. 125.

early in September 1860.[3] This was the first ever international chemistry conference, called by Kekulé so that European chemists could discuss matters of nomenclature, notation, and atomic weights. It attracted some 126 chemists. The problems facing the delegates were enormous as illustrated by the fact that values of 6/12 and 8/16 were commonly in use for the atomic weights of carbon and oxygen, respectively, and that multiple designation for the formula of a compound were common.

It was in May of 1861 that Borodin first met the acclaimed 29-year old Russian pianist Ekaterina Sergeyevna Protopopova, who was in Heidelberg recovering from tuberculosis. They attended concerts together and met other musically oriented people in a friendship that turned to love. When Ekaterina's health deteriorated and she was required to go to Italy in October for the winter months, Borodin chose to escort her there. Once in Pisa a neither of the pair wanted to separate and on visiting Sebastiano de Lucca and Paolo Tassinari at the university, Borodin found first class facilities much more attractive than those in Heidelberg. He was invited to stay and work there, which he did [5]. He even had his (by then) fiancé living in the same apartment during the 1861–1862 winter and through the spring of 1862 [5].

During the period in Pisa Alexander carried out some of the work for which he became renowned, though not completely justifiably. The laboratory was equipped with very expensive platinum retorts [19], indispensible for any work with highly corrosive materials and his intent by then was to study fluorine and its salts. He studied potassium hydrogen fluoride (KHF_2) among other fluorides and compared these salts with their simpler bases (KF), and then continued the work to organofluorine compounds (Fig. 5.4). He could perform these experiments avoiding the toxic vapours by using the retorts in the open, even in the Italian winter. This led to the preparation of benzoyl fluoride (PhCOF) from the acid chloride, the first replacement of a chlorine substituent by fluorine. The results were published in the Italian journal *Il Nuovo Cimento* [20] and *Comptes Rendus* [21] in 1862, and *Liebig's Annalen* in 1863 [22].

This was not the first synthesis of an organofluorine compound as was popularly thought. Fluoromethane (MeF) had been reported in 1835/36 by Dumas and Péligot and the higher homologue, fluoroethane, by Frémy in 1854 (Fig. 5.4c), but their studies were not well recognised. Thus, Borodin's work was popularly, but incorrectly, accredited as first synthesis of an organofluorine compound [23]. This gave him much recognition and kudos. His true fame here comes from showing that fluorides tend to form double fluorides, that sodium and potassium fluorides readily do this with a weak acid (as in Eq. 1, Fig. 5.4c), and in his use of the double salt to effect. The first 'F for C' halogen substitution (Fig. 5.4). He showed that the organic fluorides were more akin to their chloro analogues than were the inorganic fluorides to equaton 1 and the early organofluorines.

---

[3]The congress attracted chemists from Austria (7), Belgium (3), France (21), Italy (1) Mexico (1), Portugal (2), Russia (7), Spain(1), Sweden (4), Switzerland (6) and the UK (including Ireland) (17) in addition to those from Germany (56). The two attendees from Poland (War-saw) were included among the Russians. For an account of the congress see: Charles-Adolphe Wurtz (1817–1884). Account of the Sessions of the International Congress of Chemists in Karlsruhe, on 3, 4, and 5 September 1860. http://web.lemoyne.edu/~GIUNTA/karlsruhe.html#foot4. Accessed 14 June 2013.

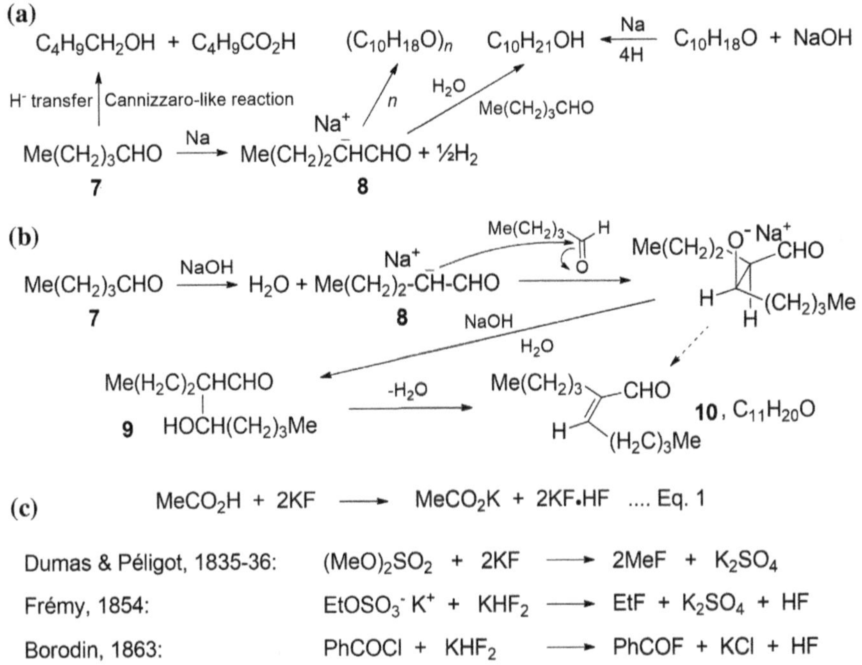

5 Alexander Porfirevich Borodin (1834–1887)

(a)

$$C_4H_9CH_2OH + C_4H_9CO_2H \quad (C_{10}H_{18}O)_n \quad C_{10}H_{21}OH \xleftarrow{\text{Na}/4H} C_{10}H_{18}O + NaOH$$

H⁻ transfer | Cannizzaro-like reaction, H₂O/n, Na⁺, Me(CH₂)₃CHO

$$Me(CH_2)_3CHO \xrightarrow{Na} Me(CH_2)_2\overset{\cdot}{C}HCHO + \tfrac{1}{2}H_2$$
7 → **8**

(b)

$$Me(CH_2)_3CHO \xrightarrow{NaOH} H_2O + Me(CH_2)_2\text{-}\overset{-}{C}H\text{-}CHO$$
7 → **8**

Me(CH₂)₃ H / O → Me(CH₂)₂ O⁻Na⁺ / CHO / H (CH₂)₃Me / H

NaOH / H₂O

Me(H₂C)₂CHCHO ← −H₂O ← Me(CH₂)₃—CHO
9 HOCH(CH₂)₃Me   H   (H₂C)₃Me
                                        10, C₁₁H₂₀O

(c)

$$MeCO_2H + 2KF \longrightarrow MeCO_2K + 2KF\cdot HF \quad \ldots \text{Eq. 1}$$

Dumas & Péligot, 1835–36: $(MeO)_2SO_2 + 2KF \longrightarrow 2MeF + K_2SO_4$

Frémy, 1854: $EtOSO_3^- K^+ + KHF_2 \longrightarrow EtF + K_2SO_4 + HF$

Borodin, 1863: $PhCOCl + KHF_2 \longrightarrow PhCOF + KCl + HF$

Fig. 5.4 The reactions of pentanal with **a** sodium, **b** sodium hydroxide and **c** the first organofluorine compounds

---

Borodin and Protopopova returned to Russia in the late summer of 1862 arriving back on September 20. However, financial difficulties prevented their marriage at that time and Ekaterina went back to the warmer Moscow to live in her mother's house and Alexander returned to St. Petersburg. Their marriage took place on April 17 the following year. At this time, Borodin was involved with the building of a new chemistry laboratory that included an apartment for the pair. Immediately on his return to the academy, Alexander became fully occupied, and in early December he was appointed as Adjunct professor [5] and began to teach organic chemistry. He was much admired by his students: a handsome man, over 1.8 m tall of a slightly oriental appearance with large deep eyes and a warm smile [4]. When working in his laboratory students felt as if they were in their own home. When Borodin was working and active in his laboratory, he never forgot music, humming tunes and talking of his new works. When he retired to his adjacent apartment the sounds of his piano could be heard by them [8].

It was shortly after his return to St. Petersburg that Borodin was introduced to Mily Balakirev, the Russian pianist, conductor and composer who inducted him to the inner circle of composers that became known as the "Mighty Handful" (see above). As the proficiency of the musicians increased, they played their works-in-progress for the group to critique [8, 24]. Borodin's life took a decisive turn from the first meeting of this group as he accepted Balakirev's challenge to continue writing music in his limited spare time under the man's tutelage. He began work on his first symphony then, the first performance of which was in 1869 under the baton of Balakirev.

Although something of an outpost of culture and academic prowess, St. Petersburg in the 1860s was an exciting place to be [24]. Borodin was starting to establish himself at the Medico-Surgical Academy of the University. He became full professor in 1864 and subsequently Zinin's successor. His friend Dimitri Mendeleev be-came a professor at the Imperial (now State) Technological Institute in the same city and, a year later in 1865, moved to the Saint Petersburg Imperial (now State) University. By 1871, Mendeleev had transformed Saint Petersburg into an internationally recognised center for chemistry research. Moreover, through his efforts, and with Borodin's support, the Russian Chemical Society was founded in 1868 by a resolution of the chemical section of the First Congress of Russian Naturalists and Physicians. It elected Zinin as its first president.

Over the years, Borodin undertook the range of consultancies expected of a professor and turned his experimental studies to what became his most concerted work. This was on the nature of aldehydes, which above all his other works ranked him with his peers. He discovered the aldol reaction almost simultaneously with Wurtz in France but was working on aldehydes before either him or August Kekulé in Germany [19]. Borodin's first published results in this area were two 1864 papers on the action of sodium on aldehydes [25, 26] These were some five years before Kekulé's first paper appeared, and Borodin felt that it intruded too much on what he then regarded as his area [19]. It was a similar occurrence with Wurtz who also did not acknowledge the Russian studies. In early March 1872, at the Russian Chemical Society meeting, Borodin disclosed three studies that gave more detail to his 1864 papers and included [19] the formation of aldol as also obtained by Wurtz [27, 28]. The issue and disputes with aldol formation lie with kudos and the scientist's place in history, not least because the reaction and the further dehydration (the aldol condensation) became so important in chemistry. So what was it Borodin did?

The first studies concerned the action of sodium metal on valeraldehyde [pentanal **7**, $CH_3(CH_2)_3CHO$] from which Borodin hoped to obtain the sodium salt, presumably **8**. However, the metal displaced hydrogen and gave a product mixture from which pentanol and pentanoic acid were isolated (Fig. 5.4a). The former could arise by reduction at the metal surface or with the acid by hydride-transfer in a Cannizzaro-like reaction. From the remaining mixture after treating with water, compounds with empirical formulae $C_{10}H_{22}O$ (15–25%) and $C_{10}H_{18}O$ (20–30%) were obtained [3]. The first of these two proved to be an alcohol while the second, although less clearly defined, was consistent analytically with two molecules of **7** combining with loss of H_2O. It can be assumed that proton abstraction from **7** gave the α-anion **8** (Fig. 5.4a) and that this reacted with a second molecule of the aldehyde to give alcohol-aldehyde (aldol) product **9** (Fig. 5.4b), which would dehydrate to the conjugated enal under aldol condensation conditions. The $C_{10}H_{22}O$ alcohol could be the diol resulting from reduction of the aldehyde function of **9**. Borodin did not establish the nature of **10** other than to state that analysis corresponded (in modern terms) to a dimeric condensation product.

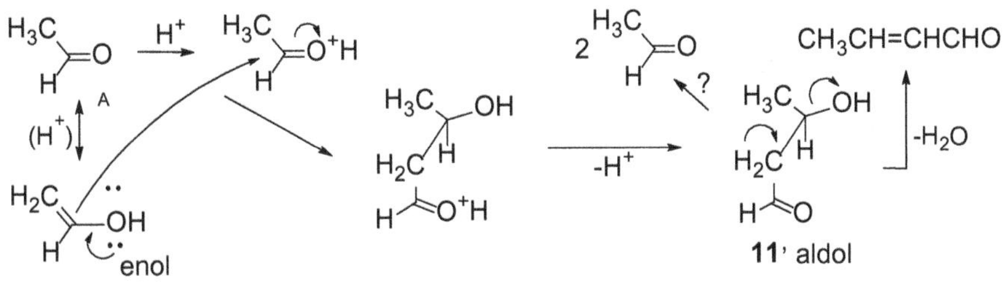

5 Alexander Porfirevich Borodin (1834–1887)

Fig. 5.5 The acid catalyzed aldol preparation

His subsequent studies involved castor oil, which on pyrolysis gives heptanal [CH$_3$(CH$_2$)$_5$CHO] and undec-10-enoic acid [CH$_2$=CH(CH$_2$)$_8$CO$_2$H]. Borodin found that this aldehyde gave a series of products similar to those from pentanal **7**. It was during this study that he examined the behaviour of acetaldehyde with HCl whereupon he obtained aldol **11** and found that it reverted to aldehyde on distillation (Fig. 5.5). The isolation of 'aldol' was at about the same time as obtained by Wurtz and represents the acid catalyzed version of the reaction. This was the subject of his third announcement at the March 1872 meeting [29]. However, Borodin conceded priority to Wurtz and granted equal priority to the earlier product obtained by Kekulé. He then retreated from the study of aldehydes. Indeed, his chemical out-put dropped and his researches reverted to the amine chemistry discussed earlier. His educational concerns came to the fore, with his professorial duties in and out of the university placing increased demands on his time. His composing remained a spare time occupation but his music became more serious as he began work on his opera Prince Igor. His final chemical works were oriented more towards medicinal chemistry and his last paper concerned improvements to a method he set up for determining the urea content of urine. He had developed hypobromite oxidation of urea (Eq. 5.1) to the stage where an accurate measurement of as little as 2.25 mL (10^{-4} M) of nitrogen could be made.

$$3NaOBr + H_2NCONH_2 \rightarrow N_2 + CO_2\ 2H_2O + 3NaBr \quad (5.1)$$

There is, however, one other study of Borodin's that deserves mention and this also concerns his medicinal chemistry. At a meeting of the Russian Chemical Society on 1 February 1871, he delivered an account of work performed by Dr. Krylov under his guidance [30]. This involved the determination of fat in heart muscles affected by fatty degeneration. Mixed with the fat they found material resembling lecithin (a generic term for any group of yellow-brownish fatty substances occurring in animal and plant tissues). They expected this to give glycerol on hydrolysis but none was found, rather a white solid that was subsequently shown to be cholesterol. Although cholesterol had been discovered in bile and in gallstones by François Poulletier de la Salle as early as 1769 and then rediscovered in 1815 and named "cholesterine" by Eugène Chevreul, its structure was not determined until 1932. The Borodin-Krylov finding of esterified cholesterol in heart lipid received little attention at the time, but it predates the recognition of harmful effects from cholesterol by approximately 40 years and is likely the first linking of the two. Borodin's dedicated work at the Medico-Surgical Academy was recognised by his promotion to academician in 1877.

Borodin's wife Ekaterina, although much affected by ill health, was a keen fighter for women's rights and she converted Alexander to the cause [1]. He was of the opinion that there should be equality of educational opportunity and became convinced that women would make good doctors. He took to opening his laboratory to students, meeting the costs himself and then, in 1872, founded the School of Medicine for Women in conjunction with Professor Rudnev and Mde. Tarnovskaya [8]. This was his proudest achievement. He inaugurated his "Course in Obstetrics" for women, and was professor of chemistry at the women's school teaching courses for some 15 years from its inception until his death. He spent less and less time in the laboratory as the demands of his women's rights campaigns and other philanthropic causes came to the fore. Because of Ekaterina's health, each summer was spent in the country with her in simple accommodations as she spent long periods in her mother's home in Moscow where the climate was kinder than in St. Petersburg. Alexander, whose heath also had never been the best [3], contracted cholera [24] in 1885 and continued to use the confinements to concentrate on his music.

The Borodins had no children of their own but sheltered and cared for many unfortunate and orphaned children. Ekaterina liked to entertain and their charity led to a home that was usually full and happy with the pair showing little concern for wealth. They adopted one girl, Liza Balaneva, who married Alexander Pavlovich Dianin, a former pupil of her father whose work was reported upon by Borodin at the 1873 meeting. Dianin gained a Ph.D. from Jenna in 1777, his medical doctorate in 1882, and joined the faculty of the Medico-Surgical Academy. He was the successor to his father-in-law and is now known for his discovery of bisphenol A and its condensation product with acetone. This latter, Dianin's compound now has importance in host-guest chemistry.

From 1872 until his death in 1887, Borodin worked on his now famous opera, Prince Igor, which was still incomplete at his death. It was finished by his musical friends Rimsky-Korsakov and Glazunov, and given its premiere in St. Petersburg on 4 November 1890 [3] Alexander Boron's death came quickly and unexpectedly. On the night of 27 November 1887, the last day of the Russian Carnival, Alexander had arranged for a fancy dress evening in one of the lecture rooms in the academy. He appeared in traditional Russian attire—a deep red woollen shirt, baggy trousers and high boots—showing his usual hospitality and was the life and soul of the party. While involved in one of the many vigorous dances he collapsed and died of a heart attack.

His esteem as a teacher was such that his students carried his coffin from the laboratory to the Tikhvin Cemetery at the Alexander Nevsky Monastery. He is buried there next to his friend Modest Mussorgsky and close to Glinka, Balakirev, Dostoyevsky, Tchaikovsky and Ruberstein. The respect in which his women students held him was shown by a silver plate on his burial casket [1]: "To the Founder, Protector and Defender of the School of Medicine for Women". The grave was surrounded by an ornate iron railing that carried at its centre a shield the circumference of which was adorned with the empirical formulae of the compounds Borodin had made [3]. Ekaterina was not in St. Petersburg at the time of Alexander's death or funeral and, when told, she became very distressed. Her health deteriorated and she died some five months later.

References

1. Profiles of Great Classical Composers. Alexander Borodin. http://www.52com-posers.com/borodin.html#man. Accessed 11 June 2013
2. Theoutsideroom (2011) The life & times of Aleksander Borodin. http://scienceandsymphonies.wordpress.com/2011/02/13/the-life-times-of-aleksander-borodin. Accessed 30 May 2013
3. Getman FH (1931) Alexander Borodin—Chemist and Musician. J Chem Educ 8:1762–1780; Getman FH (1931) Correction: Alexander Borodin—Chemist and Musician. J Chem Educ 8:2282
4. Miller R (2012) Alexander Borodin. https://myhero.com/Alexander_Borodin_06. Accessed 31 May 2013
5. Alexandr Borodin, Professional organic chemist and amateur composer. http://www.re-ocities.com/cahmn/Essays/Borodin.htm. Accessed 31 May 2013
6. Borodin A (1858) Ueber die Constitution des Hydrobenzamids und Amirins. Bull Acad St. Petersburg 17:38
7. White AD (1987) Alexander Borodin: full-time chemist, part-time musician. J Chem Educ 64:326–327
8. Friedman HB (1941) Alexander Borodin—musician and chemist. J Chem Educ 19:521–525
9. Rae ID (1989) The research in organic chemistry of Aleksandr Borodin (1833–1887). Ambix 36(3):121–137
10. Borodin A (1860) Ueber die Einwirkung des Jodäthyls auf Benzidin. Z Chem Pharm 3:533–536
11. Borodin A (1860) Ueber einige Derivate des Benzidins. Z Chem Pharm 3:641–643
12. Borodin A (1861) Ueber die Monobrombaldriansäure und Monobrombuttersäure. Z Chem Pharm 4:5–7
13. Borodin A (1861) Ueber die Wirkung des Zinkäthyl's auf zusammengesetzte Aether. Z Chem Pharm 4:8–12
14. Borodin A (1862) Ueber die Einwirkung von Zinkäthyl auf Chlorjodoform. Z Chem Pharm 5:516–519
15. Péligot E (1836) Memoire sur un acide resultant de l'action du brome sur le benzoate d'argent. Compt Rend 3:9–12
16. Borodin A (1861) Sur les dérivés monobromés des acids valérique et butyrique. Bull Soc Chim Paris 252–254
17. Borodin A (1861) Ueber Bromvaleriansäure und Brombuttersäure. Ann Chem Pharm 119:121–123
18. Hunsdiecker H, Hunsdiecker C (1942) Über den Abbau der Salze aliphatischer Säuren durch Brom. Chem Ber 75:291–297
19. Gordin MD (2006) Facing the Music: How Original was Borodin's Chemistry? J Chem Educ 83:561–565
20. Borodin A (1862) Fatti per servire alla storia de' fluoruri. Il Nuovo Cimento 15:305–314
21. Borodin A (1862) Fait pur servir à l'histoire des fluorures et préparation du fluorure de benzoyle. Compt Rend Acad Sci 55:553–556
22. Borodin A (1863) Zur Geschichte der Fluorverbindungen und über das Fluorbenzoyl. Ann Chem Pharm 126:58–62

23. Banks RE, Tatlow JC (1986) Synthesis of C–F bonds: the pioneering years, 1835–1940. J Fluorine Chem 33:71–108. (see also Ref. [9], p 127)
24. Bündler D (2001) Alexandr Borodin, Prince of Dilettantes. http://www.angelfire.com/mu-sic2/davidbundler/hero.html. Accessed 18 Jan 2019
25. Borodin A (1864) Ueber die Einwirkung des Natriums auf Valeraldehyd. Z Chem Pharm 7:353–364
26. Borodin A (1864) Ueber die Einwirkung des Natriums auf Valeraldehyd. J Prakt Chem 93:413–425
27. Wurtz A (1872) Sur an aldéhyde-alcool. Compt Rend Acad Sci 74:1361–1367
28. Wurtz A (1872) Ueber einen Aldehyd-Alkohol. J Pract Chem 113:457–464
29. Russian Chemical Society (1872) Minutes of the Russian Chemical Society meeting of 4 May 1872. Zhurnal Russkago Khimicheskago Obshchestva 4:207–209
30. Hood W (2004) Was the composer Borodin the first to link cholesterol to heart disease? http://priory.com/homol/Borodin.htm. Accessed 5 July 2013

Chapter 6
Aleksandr Mikhailovich Butlerov (1828–1886) and the Cradle of Russian Organic Chemistry

Aleksandr (Alexander) Mikhailovich Butlerov (Fig. 6.1) was born on September 15, 1828, in Chistopol near Kazan in the Tartar region of Russia [1]. The town, now named Tatarskaya, is located some 830 km east of Moscow and lies between the Volga River and its tributary the Kama. Alexander's father, Mikhail Vasilievich Butlerov, was a retired lieutenant colonel and a landowner of the district, while his mother, Sofia Alexandrovna, died shortly after he was born [2–4]. The boy spent his childhood at the home of his maternal grandparents, the Strelkovs, in the small village of Podlesnaya Shantala near Butlerovka but was brought up by his aunts with his father having significant input to his up-bringing. Mikhail's financial situation allowed him to have his son educated in the well-known Topornin private school in Kazan. It was housed in a gloomy barracks-like building on Fedorovskaya St., with a huge sign on the wall that read: *Alexander Semeonovich Topornin Boarding School for Noble Youths* [5].

After arrival at the school aged eight [3], Butlerov was subjected to the usual inauguration procedures being tormented by his fellow boarders but, according to Nazariev [5], he remained always good-natured, pleasant and polite. Butlerov was active and filled his time to the full. Initially, this manifested itself in painting but after a while, he turned to *vials, jars and bowls* [5], which one of the staff, a former army drummer nicknamed *Furious Roland*, took delight in taking away, putting the boy in a corner and/or depriving him of dinner. Butlerov persisted with his experiments while retaining his good nature and with the backing of the physics teacher. Unfortunately, on one spring evening there was a deafening explosion in the boarding school kitchen (it has been suggested this was an attempt to make gunpowder [3, 6]) and Furious Roland dragged Butlerov out by his hair with his eyebrow singed; he was followed by the older tutor accomplice who had supplied the materials for the experiments. Butlerov's crime was extreme in the view of the school board and a new punishment was required. Thus: *The criminal was taken from his dark punishment cell two or*

A previous version of this chapter was published as Halton B (2018) Aleksandr Mikhailovich Butlerov (1828–1886) and the Cradle of Russian Organic Chemistry. Chem N Z 82:46–52.

© Springer Nature Switzerland AG 2020
B. Halton, *Some Forgotten Chemists*, Perspectives on the History of Chemistry,
https://doi.org/10.1007/978-3-030-16403-4_6

Fig. 6.1 Aleksandr Butlerov (from Wikimedia Commons está disponível em português), stamp (from http://kolekzioner.net/modules/smartsection/item.php?itemid=216)

three times into the common dining hall, with a black board hanging on his chest on which great chemist *was written in large white letters* [5]. The irony of this cannot be ignored! However, Topornin, who was also the boarding school director, and a number of teachers generally treated the students well.

Following his time at the Topornin School, Alexander attended the Gymnasium in Kazan, and in 1844, at age 16, he entered the department of physics and mathematics of Kazan University. The university, founded in 1804 and the easternmost outpost of Russian higher education, had by the mid-19th century become the pre-eminent school of Russian organic chemistry despite its location and status. Butlerov studied there until 1849, initially showing interest in botany and zoology, but then followed his earlier interest in chemistry, deciding to devote his life to the science [7]. After enrolling, he worked under Professor Klaus (Claus) on the preparation of antimony but was soon attracted to Professor Zinin's organic chemistry and became an ardent pupil, even conducting experiments in his lodgings, much to the dismay of the fellow occupants. Thus, he prepared caffeine (**1**), isatin (**2**), alloxanthine (**3**) and other materials (Fig. 6.2) and took them to the university laboratory [2]. However, Professor Zinin, a mathematics graduate of Kazan who started teaching chemistry there in 1835, left Kazan for the chair at St. Petersburg Medical-Surgical Academy in 1847. Butlerov's Kazan graduation was in 1849 (a degree in natural sciences) and followed submission of his thesis, not in chemistry, but in entomology, and entitled: *Diurnal butterflies of the Volga-Ural fauna*. He had collected the material for this during his excursions around Kazan and during a trip to the steppes on the east bank of the Volga River and near the Caspian Sea in the spring and summer of 1846. Butlerov's interest in entomology continued throughout his life and he became the founder and first chief editor of the *Russian Apiculture Leaflet* in 1886 [1, 8]. Following graduation, the faculty at Kazan did not want to lose Butlerov as his abilities

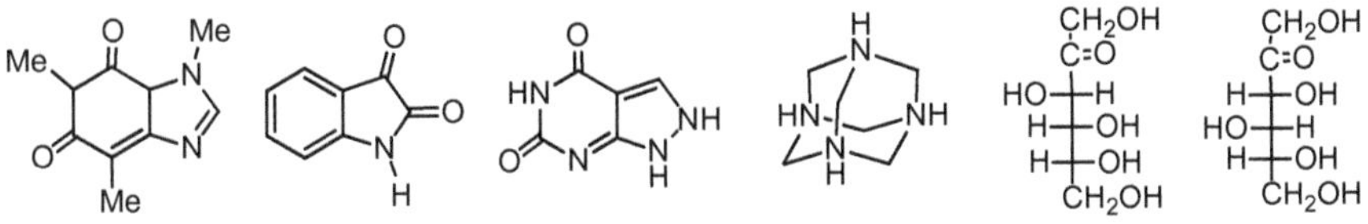
1, caffein 2, isatin 3, alloxanthine 4, hexamethylene-teramine (urotropine) D-fructose L-fructose 5, racemic fructose

Fig. 6.2 Compounds **1–5** synthesised by Butlerov

were well recognised; he was offered the only position available [2], a lectureship in physics and physical geography that he accepted and occupied from 1849 to 1850 teaching chemistry part-time until he became Klaus' personal assistant [1]. During this period, he was working on his master's thesis. The 1851 submission was entitled: *On the oxidation of organic compounds* [1], and provided a critical review of all the known information, though it also included his first experimental work on the oxidising action of osmium tetroxide (osmic acid; OsO_4) on organic compounds. He also indicated his thoughts on molecular structure saying that it is the basis for isomerism and that changes in chemical characteristics are associated with structural changes, and this at a time when such ideas had not been previously enunciated [1].

After Klaus left Kazan in 1852 for the professorship at Dorpat (the University of Tartu since Estonia became independent in 1918), Butlerov became the lecturer in chemistry and taught all of the chemistry courses and that year married Nadezhda Mikhailovna Glumilina, niece of the writer S. T. Aksakov [1]. His teaching ability immediately attracted the attention of both his students and his colleagues, although initially he taught by lecture only—work in the laboratory was not required of his students and he worked there only occasionally. Butlerov was working for his doctoral degree and had his thesis (*About essential oils*) completed in 1853. It concerned the ethereal oil of a Russian plant and the isolation of a camphor-like compound [2]. While the professors of chemistry and mineralogy approved the thesis, the professor of physics found it unsatisfactory. There was no mechanism in existence then for an independent assessment of the work and it was the decision of the Kazan Senate to allow Butlerov to submit his thesis to the University of Moscow where he defended it in 1854. He was granted the degree of Doctor of Chemistry and Physics by that university [2].

Following this, Butlerov travelled to St. Petersburg to visit Zinin and found him to be a supporter of the unitary theory of Gerhardt and Laurant, which had Laurant believe that radicals could react and were not the elements of organic chemistry as was routinely thought. On Zinin's advice, Butlerov familiarised himself with the work of Laurent and Gerhardt and became one of their passionate supporters. He was appointed an extraordinary professor of chemistry at Kazan and in 1858 an ordinary professor. By then Kazan had assumed a preeminent position in Russian organic chemistry despite its provincial location and status (Fig. 6.3). This appointment was after Butlerov made his first excursion to the science centres of Europe in 1857 together with his wife. His first stop was in Berlin where he saw and gained experience

Fig. 6.3 A. M. Butlerov with members of the Kazan Department before leaving for St. Petersburg; Kazan 1868 (courtesy of Prof Vladimir Ivanovich Galkin, Director, Bulterov Chemical Institute; enhanced for clarity). Butlerov is front row (centre) with Zinin (3rd from right). Markownikov and Zaitsev are likely 2nd and 3rd from left on back row and 19-year old E. E. Vagner is likely front row right (caption courtesy Professor A. Vedernikov)

of coal gas in the laboratory of Mitscherlich; Kazan only had spirit lamps and furnaces available. His travels through Germany encompassed most of the chemical centres including Heidelberg where he met Kekulé, at that time a privatdocent preparing his classic 1858 paper [9] in which he stated the carbon was tetratomic (valence of four) based on the simple compounds it formed with hydrogen, chlorine (in CCl_4 and CH_2Cl_2), etc. [10]. The discussions and friendship that followed impacted markedly on Butlerov's subsequent research activities. By December 1857, Alexander was working in Wurtz's laboratory in Paris. There he conducted his first serious chemical researches and discovered a new way to obtain methylene iodide (from NaOEt and I_2) and studied many methylene derivatives and their reactions. As a result, he was the first to obtain hexamethylenetetramine (urotropine; **4**) and a polymer of formaldehyde which in the presence of limewater is transformed into what was described as a sacchariferous substance—a carbohydrate containing racemic fructose **5** (Fig. 6.2) (as established by Fischer) [1], and the first complete synthesis of a carbohydrate. Whilst there he joined the Paris chemical society [3].

While most of the early advances in organic chemistry were taking place in Germany and France, the Russians were also progressing both theoretical and experimental organic chemistry. The science, as practiced in Germany and France, was brought to Russia by Alexander Butlerov and his great service to Russian chemistry was the creation of the first sustained chemistry research school in Russia. It subsequently led to the transformation of the local chemical research of the 1850-1860s, even into the1870s, to international chemical concerns and the professionalisation of the science [11]. The changes he instigated initiated professional chemistry in

Russia and were instrumental in making it the strongest scientific discipline there before 1917.

After his return from abroad in 1858, Alexander began to impart to his students the improvements he had seen in Europe to his students and installed a gas generating plant. The storage tank needed a convenient home and, as the only available space was under the floor of the laboratory, that is where it was located much to the dismay of most of the students. The provision of gas, however, made the laboratory work much easier. He then required his students to complete practical work and his noted first graduates appeared. Of these, Markovnikov, Zaitsev and Popov occupied professional chairs in universities during Alexander's lifetime. His work on the methylene series ended in 1861, when he stated the basic ideas of his theory of chemical structure and directed his experimental investigations toward the verification and support of this new theory. Summing up his research, he arrived at the theory of chemical structure, which, according to Markovnikov, he began to expound in his lectures as early as 1860. However, the beginnings of his thoughts appeared in his address to the Paris Chemical Society in 1858 and his writings of 1859. This theory was embodied in his paper *On the Chemical Structure of Substances* read at the chemical section of the Congress of German Naturalists and Physicians in Speyer on September 19, 1861 while on his second visit to European centres. At the Speyer meeting Butlerov was disturbed by the lack of agreement on a theory to explain the facts of organic chemistry [2]. It was here that he defined the concept of chemical structure. This paper was published in *Zeitschrift fur Chemie und Pharmacie* in 1861 [12] and published in the *Journal of Chemical Education* in 1971 following translation by Kluge and Larder [13]. In part, he said [13]:

> …. assuming that each chemical atom is characterised by a specific and limited quantity of chemical force [affinity], with which it participates in the formation of a substance, I would call this chemical bond or [this] capacity for the mutual union of atoms into a complex substance chemical structure.

From his definition the concept of chemical structure (a term used by Russian chemists before Butlerov, but in another sense) could only be brought forward after there had been a sufficiently clear definition of the concepts *atom*, *valence* and *interatomic bond*, as formulated in the 1858 works of Kekulé and Couper. Butlerov advanced the basic proposition of the classical theory of chemical structure by stating that the chemical nature of a compound molecule is determined by the nature of its component parts, by their quantity, and by their chemical structure [1]. On his return to Kazan his report to the university authorities said, in part [2, 14]:

> None of the ideas which I found in Western Europe seemed especially new to me. Laying aside here misplaced false modesty, I can say that these ideas and conclusions have been quite familiar in recent years in the Kazan laboratory, and they have not been considered especially original. They were developed in the general course of the work and were introduced in part into the lecture.

By the mid-1850s theoretical organic chemistry was at a crossroads. The theory of radicals was fading and the theory of types was taking its place making it possible to classify organic compounds and to indicate the types of reactions they would undergo by analogy with inorganics. The real purpose of Butlerov's 1861 paper was to try to move the chemical community from a halfway position, mixing older type and radical theories with the new theory of atomicity of elements, to the position of full structuralism [14]. His proposition, as is evident from its wording, broke with these traditional views and all the remaining proposals relating to the classical theory of chemical structure are directly or indirectly associated with that of Butlerov. The starting point of his thoughts was that for any given structural formula there existed only one compound, and for any individual compound only one formula could be written. He discarded the idea of Gerhardt who had said that the formula for a compound should vary according to its different methods of synthesis or reaction. Butlerov discarded the type theory altogether whilst Kekulé continued to use type formulae in his textbook for several years. In essence, Butlerov concluded that a structural formula should not be just an abstract image of a molecule rather than reflect its real structure. He was the first chemist to suggest that studying chemical properties of substances may lead to finding their chemical structures and vice versa [6]; he was the first to produce a model of tetrahedral carbon even thought it was irregular. Although his writings are more theoretical in nature than most of his contemporaries and his concepts those which match our accepted ideas, he went on to substantiate his theoretical views with experimental studies on aliphatic hydrocarbons and alcohols, unexpectedly obtaining *tert*-butanol as the first tertiary alcohol to be prepared. It was obtained by treating phosgene ($COCl_2$) with dimethylzinc, presumably via acetone reacting with excess of the zinc reagent. Alexander continued the study of tertiary alcohols by preparing many derivatives and improving the methods of preparation. He deserves credit for the prediction and proof of positional and skeletal isomerism, later proving, with the aid of his students, the existence of homologues. Thus, he predicted the existence of two butanes and three pentanes in 1864. Later, the expectation of isobutylene (2-methylpropene) led to the discovery of the isomeric butanes and butenes. He went on and predicted the number of isomers of the homologous hydrocarbons. He then wrote the first chemistry text based entirely on structural theory. It appeared in Russian in 1864 and in German in 1867, a little later than Kekulé's 1865 report on the structure of benzene [14]. By then, the proposals that Kekulé and Couper made simultaneously in 1858 on tetravalent carbon, and the latter's method for graphic representation, were well accepted even though they did not provide the detail that Butlerov clearly had in 1861.

By the mid-1860s, the number of chemists adopting the structural theory had increased and their way of depicting of structural formulae was that put forward by Butlerov [10]. Then, as now, the issue of priority was important and by 1867 Butlerov felt that he must stake his claim. In a footnote to his paper concerning the number of C_4H_{10} isomers [15] he pointed out that he had provided the theoretically possible number of isomers of C_5H_{12} before Friedel and Ladenburg, and went on to state that numerous chemists had made unjustifiable claims. He stated [14]:

> …the mutual chemical method of binding elementary atoms in molecules (the principle of chemical structure) will more and more be the chief basis of most chemical speculations in the newer chemistry. …and I am now obliged to assert that to me belongs an important part of the priority for complete and consequential development of this principle.

One of the chemists he named was Lothar Meyer (best known for his part in the periodic classification of the elements), who defended himself bitterly, but missed the significance of Butlerov's contributions [14]. Both Meyer's defense [16] and Butlerov's response [17] appeared in the Annalen of 1868. The latter acknowledged the major contributions of Kekulé to the theoretical and experimental branches of chemistry and this must have played its part in quietening the matter. What subsequently happened was that Butlerov's theory of chemical structure became so much a part of our science that the work of this early protagonist became and remains largely forgotten. It was in 1868, while in Nice, that Alexander's reply to Lothar was penned and during that same trip, he was advised that he had been appointed to the chair in St. Petersburg. He was also awarded the Lomonosov prize that year. In fact, it was earlier in that decade that Butlerov looked to leave Kazan. One reason for this was his unsuccessful term as rector [1]. In March 1860 he had become the last rector of Kazan University to be appointed by the Imperial Government. However, his attempts to instigate liberal changes, as well as end student abuse of individual teachers, failed and he was allowed to leave the post in August 1861. In November the following year, he became the first elected Rector of the University, but this was against his wishes. There was a struggle between groups of professors and Butlerov clashed with a university trustee, which led to his second retirement from the institution in July 1863. However, he was unhappy with the experience especially as he had not wanted the position in the first place and he began to seek a post outside of Kazan. Only the insistence of his friends, and the birth of his only child, a son, Vladimir Aleksandrovich, in April 1864 stopped him from departing immediately [1].

Butlerov's transfer to St. Petersburg University as Professor of Chemistry ran from 1868 until 1885, when he was required to retire on pension. Despite this, he continued to give special lecture courses. Much of his time at St. Petersburg was spent defending and substantiating the theory of chemical structure from his university colleagues Menschutkin (his successor) and, to a lesser extent, Mendeleev [1]. Butlerov felt strongly that chemists not only had the right, but also the responsibility, to speak of molecules and atoms as though they existed and, in so doing, preserved the conviction that this belief would not become a baseless abstraction. Whether Butlerov is the father of chemical structural theory, as implied by many Russian authors, or whether he simply played a major role in its development and acceptance is debatable [18]. Nonetheless, a prominent group of Russian organic chemists worked with Butlerov at St. Petersburg as in Kazan (Fig. 6.4). Of these, A. E. Favorskii is the most noted (see below), though I. L. Kondakov deserves mention as he and his co-workers achieved remarkable success as he was the first to produce synthetic rubber in 1901 when Head of Chemistry at Tartu University (1894–1918). It should also be noted that Egor Egoroviè Vagner, a graduate of Markownikov, also worked in the Butlerov laboratory. Vagner (born in Kazan) subsequently gained a chair in Warsaw and published under the German translation of his name, Georg Wagner [19]; he discovered that α-pinene rearranged to bornyl chloride in 1899 and is the Wagner of the Wagner-Meerwein rearrangement.

Fig. 6.4 The noted Butlerov students from left: Marknovnikov, Zaitsev, Kondakov, and Favorskii (all in the public domain from Wikimedia)

Butlerov's outstanding characteristic as an instructor was that he taught by example and his students could always observe what he was doing and how he was doing it; he always answered questions and was never dismissive of his students [1]. He was a staunch advocate of higher education for women and from 1878 organised university lecture and laboratory courses in chemistry for them at St. Petersburg. In addition, Butlerov delivered there, as he had earlier done in Kazan, a large number of public lectures, most of which had a chemical and/or technical basis. By the end of the 1870s, organic chemistry in Russia had reached the stature needed to make it unnecessary for a student to leave the country for advanced training in the discipline. By then there were vibrant schools of chemistry headed by Russian and Russian-trained professors. Thus, St. Petersburg was served by Zinin, Butlerov, Menshutkin and Borodin,[1] Moscow had Markovnikov, and Kazan, Zaitsev.

Although Butlerov did nothing to improve the ways and procedures for identifying or synthesising new compounds with particular structures, he experimented to prove his theory. He was able to formulate isomerism rules from the theory, and chemists were able to predict the existence of the four valeric acids ($C_4H_9CO_2H$), three of which were synthesised outside Russia in 1871, while the last isomer was synthesised in Butlerov's own laboratory in 1872—verification not only by the proponents but also independently. At the very least, Butlerov must be accepted as *one of the main philosophers of the structural theory* [3].

Butlerov also spent considerable time on agriculture, horticulture and beekeeping. He wrote a book on beekeeping in 1871 entitled: *Bee, its life and the main rules of sensible beekeeping*. It went to ten editions before the Russian Revolution and was the practical handbook of its time [4]. In his later life in the 1860s he grew tea in the Caucasus [6] where the tea plant flourished on the western lowlands of Transcaucasia (modern Georgia, Armenia, and Azerbaijan), notably at Soukhourn [20].

In the spring of 1886 Butlerov's health deteriorated and, though not thought serious, he ended his work and returned to his estate in Butlerovka on the banks of the Kama River. He died there on August 17, aged 58. He was buried in the family chapel in the rural cemetery. A statue of him was erected in Kazan in 1978 and the successes

---

[1] For Borodin, see Chap. 5.

of the Kazan chemists contributed to the organisation of a Research Chemical Institute named after him in 1929 at Kazan University. As a result of the reorganisation of Soviet Universities in 1933, the Department of Chemistry in Ka-zan University was reopened. Then, in 2003 the Alexander Butlerov Institute of Chemistry was established when the Research Chemical Institute and the Chemical Faculty of Kazan State University merged.

6.1 Brief Commentaries on Butlerov's Noted Students in the Cradle of Russian Organic Chemistry

6.1.1 Vladimir Vassil'evich Markovnikov (1838–1904)

Vladimir Vassil'evich Markovnikov was born on December 22, 1838, in Nizhny Novgorod about 400 km east of Kazan. He studied under Butlerov in Kazan (1860 graduate) and St. Petersburg making his most noted contribution to organic chemistry in 1869, namely the Markovnikov Rule, which predicts the regiochemistry of addition reactions. He was a lecturer at Kazan in 1862, succeeded Butlerov in 1869, gained a chair in Odessa in 1871 then transferred to the chair in Moscow in 1873. His further contributions to the field of organic chemistry were his discovery of the naphthalenes in the early 1860s, the cyclobutanes in 1879 and the cycloheptanes in 1889. He died in Moscow on February 11, 1904. He was a founder of the Russian Chemical Society in 1868.

6.1.2 Alexandr Nikiforovich Popov (1840–1881)

Popov was born in about 1840 in Vitebsk in Belarus. He studied at Kazan under Butlerov graduating with his Ph.D. in 1865. In 1869 he graduated with a D.Sc. degree from Kazan and was appointed to Warsaw University on its inauguration that year where he remained until his untimely death on August 6, 1881. He was known for the Popova Rule which came from his study of the oxidation of organic compounds by a chromium mixture; the rule allows methods of determining the chemical structure of ketones, acids, alcohols, and hydrocarbons.

6.1.3 Aleksandr Mikhaylovich Zaitsev (1841–1910)

Zaitsev graduated with a degree in economic science in 1862 but had to study chemistry as well, to which he was converted by Butlerov. At the time of his graduation he was no longer an economist, but a committed chemist. Not complying with Russian traditions, Zaitsev moved to Germany and then Paris where he studied under several of the noted chemists. His return to Kazan was not simple, but in 1869 he was appointed to the chair in Chemical Technology while Markownikov held the chair of chemistry; there was little love lost between the pair. Whilst at Kazan, Zaitsev enun-ciated his empirical rule for elimination, the Zaitsev (or Saytzeff) rule which states that the most substituted product will be the most stable, and therefore the most favoured. He remained at Kazan and had as students Egor Egorevich Vagner (Georg Wagner, 1849–1903) and Sergei Nikolaevich Reformatskii (1860–1934). Zaitsev died in Kazan on September 1, 1910.

6.1.4 Ivan Laverntievich Kondakov (1857–1931)

Kondakov was born on October 8, 1857 in Viluisk in the Yakutsk territory of Eastern Siberia. He studied at St. Petersburg University under Butlerov then transferred to Warsaw University in 1886. In 1895 he moved to Tartu University in Estonia where he discovered the polymerisation of 2,3-dimethylbutadiene (1899). He synthesised synthetic rubber (methyl rubber) in 1901 and his procedure was used in Germany in 1916. He also studied the rearrangement(s) of 2-methylbut-2-ene seeking a diene product. He died in the village of Elva in Estonia in 1931.

6.1.5 Alexey Yevgrafovich Favorskii (1860–1945)

Favorskii was born in Pavlovo, Nizhny Novgorod on March 3, 1860. He studied chemistry at the University of St. Petersburg from 1878 to 1882 under Butlerov, where he then stayed for several years. In 1891, he was appointed lecturer, gained his Ph.D. in 1895 and became professor for technical chemistry. His discovery of the Favorskii rearrangement in 1894 (the rearrangement of an α-haloketone to a cyclopropanone and then a carboxylic acid) and the Favorskii reaction (the nucleophilic attack of an acetylide on a carbonyl group) between 1900 and 1905 are his most noted works. He was at the (new) inorganic department from 1897, and served as its director from 1934 to 1937. For his improvement of the production of synthetic rubber, Favorskii was awarded the Stalin Prize in 1941. He died on August 8, 1945 and is buried at the Volkovskoe Orthodox Cemetery in St. Petersburg.

References

1. Encyclopaedia.com (2008) Butlerov, Aleksandr Mikhailovich. http://www.encyclope-dia.com/people/science-and-technology/chemistry-biographies/aleksandr-mikhailovich-butlerov. Accessed 8 May 2017

References

2. Leicester HM (1940) Alexander Mikhaïlovich Butlerov. J Chem Educ 17:203–209
3. Konovalov AI (2011) The Butlerov theory of the structure of organic compounds, Report on the international congress on organic chemistry, Kazan, 19 Sept 2011
4. Arbuzov RA (1978) 150th anniversary of the birth of A.M. Butlerov. Russ Chem Bull 27:1791–1794. (Translated from Izvestiya Akad Nauk SSSR Ser Khim 1978, 2035–2039)
5. Nazariev V, Bogdan N (trans) Life and people of times past. http://www.ninabog-dan.com/uploads/Life_and_People_of_Times_Past_edited.pdf. Accessed 10 May 2017
6. Kurbanova DM (2016) Butlerov, Alexander Mikhailovich. http://www.studfiles.ru/pre-view/6055263. Accessed 11 May 2017
7. Kizilova A. Alexander Butlerov. http://russia-ic.com/people/general/b/109. Accessed 12 May 2017
8. Lewis DE (2012) Early Russian organic chemists and their legacy. Springer Briefs in molecular science: history of chemistry. Springer, Heidelberg
9. Kekulé A (1858) Ueber die Constitution und die Metamorphosen der chemisehen Verbindungen nnd iiber die chemische Natur des Kahlenstoffs. Ann Chem Pharm 106:129–159
10. Hiebert EN (1959) The experimental basis of Kekule's valence theory. J Chem Educ 36:320–327
11. Brooks NM (1998) Alexander Butlerov and the professionalisation of science in Russia. Russ Rev 57:10–24
12. Butlerov AM (1861) Einiges über die chemische Structur der Körper. Z Chem Pharm 4:549–560
13. Kluge FF, Larder DF (trans) (1971) A. M. Butlerov. On the chemical structure of substances. J Chem Educ 48:289–291
14. Leicester HM (1959) Contributions of Butlerov to the development of structural theory. J Chem Educ 36:328–329
15. Butlerow A (1867) Ueber die Derivate von Trimethylcarbinol (von tertiärem Pseudobutylalkohol). Isomerie der gesättigten Kohlenwasserstoffe C_4H_{10} und der Butylene C_4H_8. Isobutylalkohol (der primäre Pseudobutylalkohol oder Pseudopropylcarbinol). Ann Chem Pharm 144:1–32
16. Meyer L (1868) Zur Abwehr. Ann Chem Pharm 145:124–127
17. Butlerow A (1868) Eine Antwort. Ann Chem Pharm 146:260–263
18. Rocke AJ (1993) The quiet revolution: Hermann Kolbe and the science of organic chemistry. University of California Press, Berkeley, pp 257–264
19. Morris P. Biographies of chemists. http://www.chem.qmul.ac.uk/rschg/biog.html. Accessed 19 May 2017
20. Anon (1895) Tea cultivation in the Caucasus. Bull Misc Inf R Bot Gard 99:58–61

Chapter 7
Heinrich Caro (1834–1910)

Heinrich Caro (Fig. 7.1) was born in the town of Posen in Prussia (now Poznan, Poland) on February 14, 1834, son of Simon and Amelia (née Schitzer) Caro. His parents were of Jewish decent, his father's family having emigrated from Portugal to the town of Glogau in Silesia. His mother came from a jeweller family who had moved from Amsterdam to Wroclaw (Breslau) in Silesia. After moving to Posen, Heinrich's father and grandfather established a grain merchant company. Of Heinrich's five siblings only his two-year elder brother Julius survived. Shortly after Heinrich started school, the family moved to Berlin via Breslau, his mother's hometown, in the hope of a better education and life for their family [1].

Once in Berlin, Heinrich attended the Kölnische Regalgymnasium for 10 years from the autumn of 1842. His youth in Berlin was centred on music, adventure and drama. His mother, who had a strong interest in opera, engendered in him a significant interest in music and had the boy learn German folk tunes and listen to military bands [2]. He was part of an informal drama circle, after one of whom (Anna Peters nicknamed rosenfingerigne, Eos) he subsequently named his dye eosin (eos Greek; goddess of the dawn). He became artistically oriented and was labelled as a dreamer by a colleague in his BASF days. The fourteen year old Caro saw first-hand the revolution of 1848 that sought liberal reforms; his home was in a building that also housed the much disliked state censor and mayor of Berlin. The street battles and their aftermath impacted significantly on him. According to Travis [2] the impact of these events allowed Caro to retreat into a dream world of his own and play out the adventures of his boyhood heroes.

The Regalgymnasium had a chemical laboratory and the young Caro became fascinated with the experimental work as he saw the subject offering much advantage to all branches of the trades. Prior to graduating, he was performing experiments at home that he had done in the lab and others he obtained from textbooks. From there he moved for trade training at the Royal Commercial Institute becoming a colourist

---

A previous version of this chapter was published as Halton B (2015) Heinrich Caro (1834–1910). Chem N Z 79:200–206.

Fig. 7.1 Caro, from BASF AG self-scanned in Wikipedia and an example of his indigo with cakes by David Strone (http://en.wikipedia.org/wiki/Indigo_dye#/media/File:In-digo_cake.jpg)

in cotton dyeing and printing, and taking chemistry lessons at the University of Berlin. In doing so, he likely had some training and input from the Berlin textile manufacturer Benjamin Lieberman, the largest calico-manufacturer in Germany and father of chemist Carl Liebermann with whom he subsequently collaborated. As an outstanding student graduating in 1855, Caro gained a position as colourist with calico printer Troost in Mülheim an der Ruhr, where natural dyes and secret recipes were still in use.

Before the mid-1850s, all commercial dyes were obtained from natural products and the experts in their application and improvement were textile colourists. Their job was to analyse the (mainly) vegetable-based extracts, e.g. the blue from indigo and the red from the madder plant root, and improve dye fastness. This required the application of then known chemistry involving simple chemical analyses of the carbon-containing vegetable products and of certain metal compounds, the mordants that fixed the dye to the fabric. These often led to distinct colours.

This was the era into which Caro entered the textile industry. He soon demonstrated the power of scientific solutions to certain production problems and was sent to England in 1857 to learn the up-to-date dyeing techniques including advances in the use of steam. This was the year after William Perkin's discovery of mauve from aniline. Caro gained the needed information on the developments in England where he spent some time with the firm of Roberts, Dale & Co. in Manchester; he had met John Dale in Germany whilst a student [3]. On return to Mülheim and Troost he was required to undertake military service and on discharge in 1858 Caro joined the new calico printing company, Louisenthaler Actiengesellschaft in Mülheim, but he stayed there for less than a year. He was ambitious and restless, and felt he could make a name for himself more easily in England than in his home country.

Back in England, Caro soon gained a permanent position with Roberts, Dale & Co. (at their Cornbrook Mill at Hulme in Manchester) and he stayed there until

1866. He was given the responsibility for quality control and for providing calico printing clients with the recipes for natural colours that included the madder extract, alizarin. At that time, South Lancashire was increasingly becoming the centre of British chemical industry. Large-scale coal-tar production came to the fore from about 1859 and the development of synthetic dye manufacture gave a new aspect to chemical industry closer to academic chemistry (even pharmacy) than the then more traditional manufacture of acids and alkalis [4]. By the spring of 1859, mauve (mauveine) manufactured by Perkin & Sons in London was at the forefront of fashion. This led to the coal-tar hydrocarbons benzene and naphthalene being nitrated and reduced to amines that were, in turn, treated with a vast range of oxidising agents in the hope of providing new synthetic colourants in what is now recognised as the aniline dye industry. It was this environment that saw Caro back in Manchester where he became a key figure in the development of 19th century synthetic dyestuffs [4]. Roberts, Dale & Co. decided to become a significant player in this market and used to the full its German employees Caro, Martius and Leonhardt, and consultants Carl Schorlemmer (Owens College, Manchester) and Peter Griess (Allsopp's Brewery, Burton on Trent).

Working in his own time, Caro undertook a series of experimental studies with aniline in the hope that something practical might ensue. Inspired by Perkin [4], Caro subjected aniline to a variety of oxidising agents and found that copper sulfate and sodium chloride provided Perkin's mauve more efficiently than the potassium permanganate used by him. Together with John Dale the process was patented in May 1860, subsequently to go into production, with Caro made a partner in the firm to work the patent. He was always careful to ensure that his mauve was suitable to his consumers and over the ensuing years, improved versions were available from both Roberts, Dale & Co. and Perkin & Sons. The Manchester dye provided Perkin with his sole competitor of significance and it was so successful there that Heinrich Caro adapted and mechanised the equipment for the manufacture of nitrobenzene and aniline to allow the company to expand. His understanding of the needs of the calico printer and his after-sales service in his work for Dale, Roberts & Co. was to the extent that he was given charge of marketing their full range of products including the mauve, and it was left to him to set the selling price of their dyes in Europe. As we will see below, this contrasts with his reputation at BASF.

It is worthy of mention that it was only in 1858 that Kekulé and Couper had advanced the concept of tetravalent carbon and that Kekulé's historic six-membered ring structure of benzene did not appear until 1865. Thus, it is not surprising that the nature of the Perkin mauve (mauveine) was not settled until the 20th century, in fact only in 1994 when the presence of four phenazinium salts [3-amino-2-methyl-5-aryl-7-(p-tolyl)amino derivatives; mauvine A, B, B2 and C] were identified (Fig. 7.2) [5]. The presence of the o- and p-toluidine moieties in mauvine stem from the impurities present in the aniline of the late 1850s and 1860s and they are critical to the synthesis. The proportion of the components in mauvine depended upon the quantities of the impurities, but mauvine A (from aniline:o-toluidine:p-toluidine in a ratio of 2:1:1) always dominated; now, a total of 12 components have been identified [6].

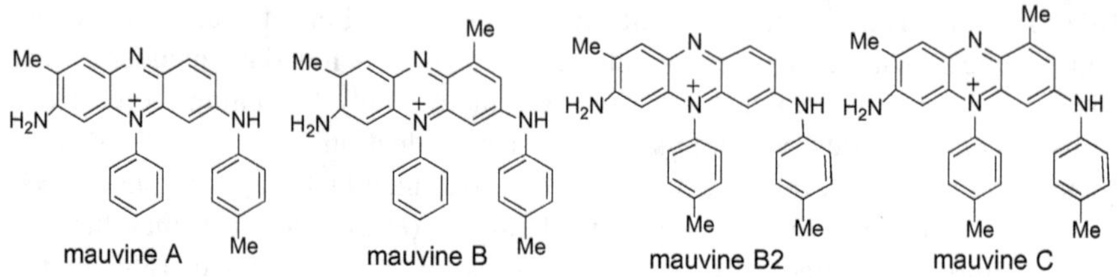

Fig. 7.2 The structures of mauvine present in the Perkin and Caro products; the dye was commonly the acetate salt

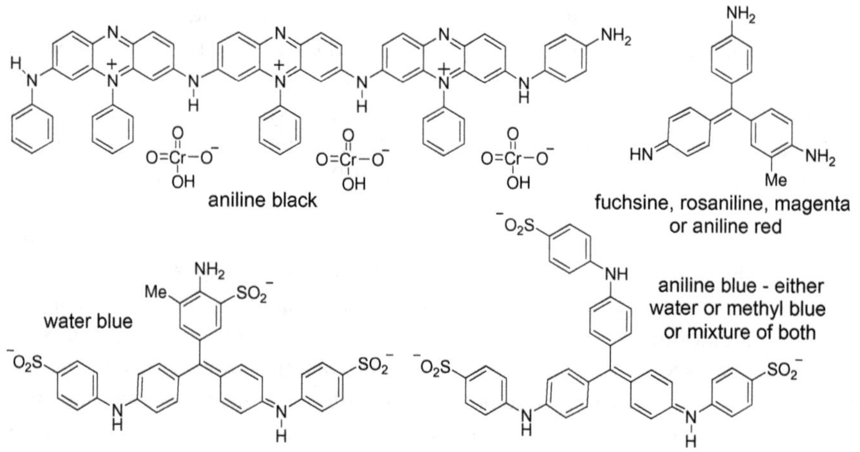

Fig. 7.3 Some examples of the Caro dyes

The Caro mauve process was not entirely straightforward as it left a black residue [2]. However, this residue provided a first-rate fast black (aniline black) for calico printers from approx. 1862 (Fig. 7.3). John Lightfoot of Accrington had first provided this compound but his method employed the use of aniline and copper salts directly in the printing process. The effect of the copper was slowly to corrode the steel rollers of the printing machines in England, but the procedure worked well with the wooden rollers used in continental Europe. The Caro product had no such undesirable effect as the dye had been independently produced and it gave the company significant advantage.

By the end of 1862 both Perkin & Sons and Roberts, Dale & Co. had converted mauve into other colours and in late January 1864 Dale and Caro patented their after-treatment of magenta, mauve, and other aniline colours with acidic acrolein (propenal; $CH_2 = CH-CHO$) that provided purple, blue, greens and other colours. These widened the range available and began to reduce the popularity of mauve [7].

A. W. von Hofmann at the Royal College of Chemistry in London, and F.-E. Verguin in Lyon, independently showed that (impure) aniline could give a new red dye, sometimes known as aniline red (fuchsine in France, rosaniline or magenta in

England) [3, 4]. It became the most sought after colour after mauve. In his synthesis, Hofmann used carbon tetrachloride while Verguin, who patented the process, used stannic chloride. Subsequently, Henry Medlock, a former Hofmann student, patented an improved process using arsenic acid as the oxidising agent. Although Caro could prepare the new dye using lead nitrate, it was of no better quality than that from the less expensive arsenic route. Once again, however, the manufacturing process was substantially improved at Roberts, Dale & Co. thanks to another Caro contact and subsequent employee, August Leonhardt. His Manchester procedure was so successful that Leonhardt was even able to promote it after his return to Germany in 1867.

Another new aniline dye of significance was aniline blue, discovered in France by Charles Girard George de Laire in 1860 [3, 4]. It was obtained only with an excess of aniline in the mixture that gave aniline red. It was patented and became popular. Today it is recognised as being comprised of water blue, methyl blue or of a mixture of the two compounds (Fig. 7.3). Although Caro and the company were interested in the blue dye, patents kept them out of manufacture. Nonetheless, the Roberts, Dale & Co. consultant Carl Schorlemmer (Owens College) showed that the use of inorganic acid in the Girard de Laire process was inferior to the use of benzoic acid in a newer process devised by Wanklyn and Eisenlohr. Caro and Wanklyn published their results in the area [8], subsequently shown to have incorrect formulation, but published a few weeks before Hofmann's ground breaking paper on the composition of aniline blue as a triphenylated derivative [9–11]. This latter paper showed the blue dye to carry 18 carbons and 12 hydrogen atoms more than the red dye and this was easily explained by the presence of three additional benzenoid rings. It was this work that opened the way for interaction between the industrial and academic chemists, though it was Caro and his subsequent move that really laid the foundation for technology transfer as we now call it. The triphenyl dyes studied by Hofmann had also aroused Caro's attention and the study of these led to his subsequent academic–industry career.

Carl Martius joined Roberts, Dale & Co. in 1863 from Hofmann's London laboratory and soon had a new brown dye available that was named Manchester brown (or phenyl brown) in Britain, but Bismark or Martius brown in Germany (Fig. 7.4a) [3]. It came from the reaction of m-diaminobenzene with nitrous acid and shortly thereafter aniline yellow was provided from p-diaminobenzene and aniline. It was from the consultancy work of Peter Griess, done in his own time while working at the Allsopp's Brewery [3] that the wherewithal for the reaction of an aromatic amine with nitrous acid came. The now well-known diazotisation process that provided so many industrial dyes originates from the 1858 publication by Griess [12, 13]. Although their compositions were unknown, Manchester brown and aniline yellow were patented in 1863. That year Caro obtained the blue colour induline from reaction of aminoazobenzene with aniline and its hydrochloride at high temperature. It was a long-lived product and one of his major discoveries, again emanating from the exploratory work of Griess. The actual structure of the product is dependent upon the temperature and time period of the reaction and now several indulines are known; the most common are the 6B and 3B varieties as shown by Kehrmann's authentic syntheses (Fig. 7.4b) [14–17].

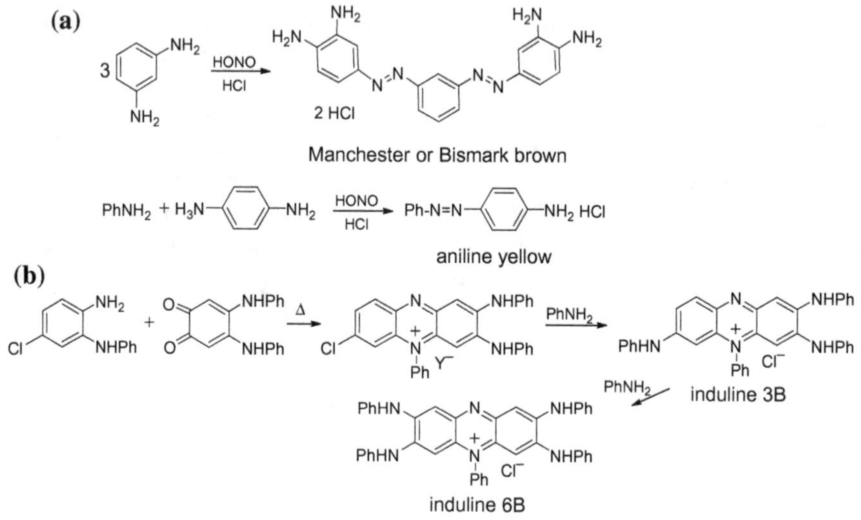

Fig. 7.4 Synthesis of **a** Manchester brown and anline yellow, **b** induline 3B and 6B

It was Caro's training as a colourist in the printing industry that enabled him to fully appreciate the technical challenges of dye application, and through this he facilitated the smooth transition from natural to artificial colourants. The various investigations and inventions on aniline and its derivatives carried the hallmarks of a new high-technology industry in the making. By the mid-1860s mauve had been displaced by the newer aniline dyes. However, mauve saw extensive use in wallpaper and paper printing and, through Caro's efforts, in the printing of postage stamps, probably its last commercial application [18].

Caro developed respiratory problems during his time in Manchester and given this, and the better prospects that had become available in his home country, he resigned his position and returned to Germany in late 1866. But this was not before he married. He first met his future wife Edith Sarah Eaton in 1862 when she was twenty years old. They married on October 31, 1866, at Saint Saviour's Church in Chorlton-on-Metlock, south of Manchester city. The couple had a honeymoon on the Isle of Wight before leaving England for a brief sojourn in Berlin before settling in Heidelberg, where Heinrich worked in Bunsen's laboratory. They had three sons and four daughters.

During his time in Manchester Caro maintained contact with his German contemporaries and extended his circle of contacts so that while working in Bunsen's laboratory he was able to develop further his entrepreneurial skills. He became a consultant to the company Badische Anilin-und Sodafabrik formed in April 1865 by Friedrich Engelhorn, a trained goldsmith who became an entrepreneur in the gas lighting industry establishing a company in Mannheim. Engelhorn saw potential use in his major waste product, coal tar, as a dye precursor. His initial objective was to have a factory sited on the Mannheim side of the Rhine but attempts to purchase a site

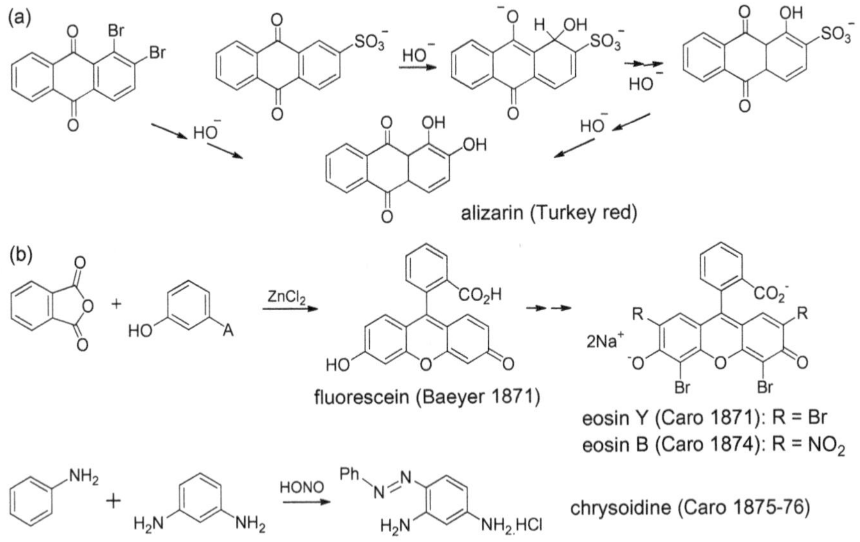

Fig. 7.5 Syntheses of **a** alizarin and **b** fluorescine and eosin Y and B, and chrysoidine

fell through and so he set up BASF on the opposite side of the river in Ludwigshafen. Almost immediately, he began to construct housing for his workers, as the town became the fastest growing in Germany. The coal tar dyes that the company manufactured were aniline red and its derivatives. However, they disappointed, being neither colourfast nor lightfast and these factors (amongst others) led to Heinrich Caro's consultancy work [19].

By 1868, it was clear that the traditional natural dyes could no longer supply the growing demands of the textile industry and the need for intensive chemical research by BASF became obvious. Thus, it was that Engelhorn, the Head of BASF, appointed Heinrich Caro as BASF's technical director and head of the test laboratory in 1868. His laboratory at that time was in Mannheim until the building was sold and the facility moved to a lab just inside the main gate of the Ludwigshafen site. Caro became the company's first Research Director in the mid-1870s. His initial work was when the company was seeking to commercialise the production of synthetic alizarin (Turkey red, Fig. 7.5), which along with indigo was the most important dyestuff. That year, Carl Graebe and Carl (Theodore) Liebermann, each studying for his habilitation, showed that alizarin could be degraded to anthrcene and the following year they announced its synthesis from anthracene. It was the first laboratory preparation of a natural dye [20, 21]. Although they patented their process [22], Graebe and Liebermann sold their rights to BASF because the cost of the bromine needed to give the essential dibromoanthraquinone for the synthesis (Fig. 7.5) was too high for commercialisation.

Caro's first job at BASF was to find a cost-effective modification of the synthesis and this he did by sulfonating anthraquinone and then treating the resultant acid with base as shown in Fig. 7.5 [23]. The trio then patented the process for BASF, filing just one day before Perkin filed his patent for essentially the same process. Caro negotiated settlement, the outcome of which was a Perkin-BASF cartel that divided the international market for this important dye; Perkin had the UK and BASF Europe and the US. Commercialisation of the alizarin process was not easy and Carl Glaser, who had been taken on from Kekulé's Bonn laboratory following his habilitation in 1868, assisted Caro in obtaining a useful process. The difficulties revolved around the purification of the starting materials, but their efforts gave synthetic alizarin as the first major success of the BASF. Glaser continued to be associated with Caro for most of the time both were with BASF (see below). The alizarin process has importance because its inventors and investigators gave tremendous support to the benzene-ring theory from their extensive studies. This led to the now traditional structural formulae for naphthalene and anthracene. The very early adoption of the benzene-ring theory by German chemists, particularly those involved in solving industrial problems, initiated the long period of competitive advantage held by the German dye and chemical industry.

Following the alizarin work, Caro soon developed a fluorescent dye that he named eosin. Its origin came from synthetic work by von Baeyer in Munich that gave fluorescein in 1871. Caro took the Baeyer compound and brominated it to give his new eosin dye the same year—a tetrabrominated fluorescein (Fig. 7.5). Some three years later he had modified his process and produced a dinitrodibromo analogue now recognised as eosin Y (yellowish) and eosin B (bluish), respectively (Fig. 7.5). The following year (1874) saw the discovery of chrysoidine, an orange azo dye. Although there seems to be some debate on whether it was Caro alone or in collaboration with Otto Witt (working at the Star Chemical Works in Brentford, Middlesex) who made the discovery, it was Witt who commercialised the dye although he did not patent it because the structure was not known. It proved to be the product of simple diazo coupling of aniline to m-diaminobenzene following the Griess studies and it marked an era in the azo dye industry (Fig. 7.5).Caro's next success came in 1876 and marks what has to be his major contribution to chemistry, namely the discovery of methylene blue [2,7-bis(dimethyla-mino)phenothiaz-5-ium chloride; Fig. 7.6]. It comes as dark green crystals or crystalline powder that has a bronze-like lustre but its solutions in water or alcohol have a deep blue colour. It was prepared by Caro from oxidation of 4-amino-dimethylaniline with ferric chloride in the presence of hydrogen sulfide dissolved in hydrochloric acid. It was patented [24] by him for BASF in 1876 and was the first dye from coal tar patented in Germany. Apart from this, methylene blue has been described bas the first fully synthetic drug used in medicine [24] and was pioneered for treatment of malaria. It continued to be used in WWII, where it was not well liked by soldiers who stated: Even at the loo, we see, we pee, navy blue. Of note is that the production of blue urine monitored the compliance of psychiatric patients with medication regimes. Antimalarial use of the blue dye has recently been revived but led to interest from as early as the 1890s to the present day from the drug's antidepressant and other psychotropic effects. It became the lead compound in research that led to the discovery of chlorpromazine, a dopamine antagonist of the typical anti-psychotic class of drugs used to treat schizophrenia.

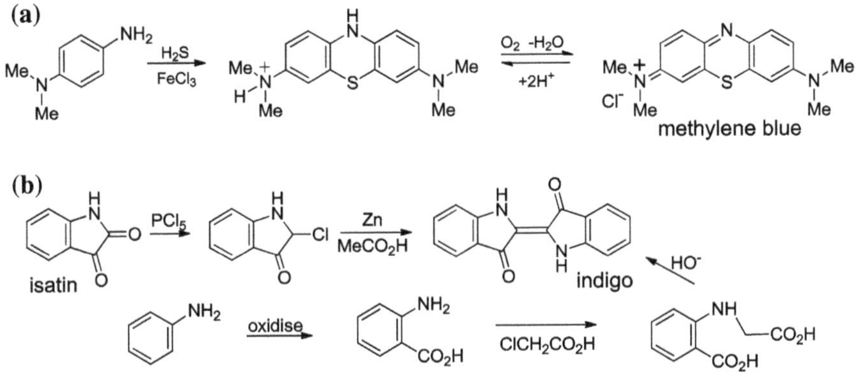

Fig. 7.6 Syntheses of **a** methylene blue and **b** indigo

The year of 1877 saw Caro develop Fast red A, which was followed by malachite green and rosolic acid [25, 26]. That year he not only cofounded the association for the interests of the German chemical industry (now the German Chemical Industries Association) but he also gained his first honorary doctorate (The University of Munich). By this time, he had become friendly with Adolph Baeyer in Munich and both of them began collaborating in the search for a suitable industrial synthesis of the natural dye indigo that Baeyer had prepared in 1870. The work was never successful but lasted for many years.

Baeyer recognised the potential of synthetic dyes and from 1865 he searched for a laboratory synthesis of natural indigo. It was extracted from tropical plants in the indigofera genus cultivated extensively in India and demand for it was high. Baeyer's success followed his establishment of the structure in 1870 and involved treating isatin as depicted in Fig. 7.6b. However, isatin was too expensive a starting material for commercial synthesis and Caro came to an agreement with Baeyer to seek an appropriate industrial process collaboratively. BASF purchased Baeyer's indigo patent for a substantial sum but nothing that could be used industrially was ever found, despite a further synthesis by Baeyer. The breakthrough came in 1890 when Karl Heumann in Zurich found a way of making indigo from aniline (Fig. 7.6). A lucky accident at BASF involving a broken thermometer then revealed that mercury was a catalyst for a key part of the synthesis and the German company started production of synthetic indigo in 1897. Johannes Pfleger then discovered a better route to synthetic indigo that involved converting aniline into N-phenylglycine, which on fusion with NaNH$_2$ yielded glyoxyl that oxidises in air to indigo; this was used by BASF's competitor, Hoechst. This dye for blue jeans remains in high demand and is legendary.

By 1880, Heinrich Caro had become BASF's expert in patent issues and dealt with all such applications. He went on to become a leading spokesman for the German chemical industry, helping to develop patent law to protect chemical inventions. In 1884, he advanced to the BASF Board of Executive Directors. His personal nuances working for BASF have been provided by Carl Glaser, and described in some detail by Anthony Travis [2]. To this author it is clear that Caro and Glaser had very different personalities and their working together for many years must have exacerbated matters between them. Glaser was the (perhaps) typical industrialist who worked regular hours seeing his goal as evolving industrial scale procedures for the company. In contrast, Caro appears to have evolved into an academic seeking new chemicals and valuable end products. He was an expert in the laboratory regularly singing when at work, but he became irregular in his work hours and this frustrated the systematic Glaser. Nonetheless, Caro provided BASF in its early years with products and procedures that positioned the organisation as a major player in dyestuffs. His laboratory became recognised as the first industrial research facility. He had 22 chemists pass through his lab and was responsible for some 22-dye patents, another eight from other related labs, and an additional 19 company patents that he was involved with. When Caro left BASF there were some 18 small laboratories on the Ludwigshafen site in addition to the newly opened Central Research Laboratory.

Glaser lays much of Caro's success to the persuasion that Englehorn was able to exert on his chief chemist until the latter's departure from the company in 1885. The acclaim that Caro gained has to have substance and, to this author, it seems that Caro's great strength was to see a valuable chemical and then modify the discoverer's synthesis of it to provide a commercially viable process. With the exception of methylene blue and Caro's acid (see below) most of his successes came from such effort, probably with much of his work carried out in his home laboratory in Mannheim. There is little doubt that Caro was responsible for more of Germany's successes in the dye industry than any other individual was. Heinrich Caro left BASF at the end of 1889, not long after the Central Research Laboratory that he had designed opened; it became a formal business unit. Heinrich Caro transferred to the BASF Supervisory Board for the rest of his life.

Caro's contribution to the 1890 Kekulé 25th benzene anniversary did not appear until 1891 and was his history of the dye industry over its formative years. This work, now regarded as an epic [27], provides the most complete coverage of the era. It does this by Caro creating an imaginary dyestuffs manufacturer and describing in detail the operations and developments involved.

The manufacture of alizarin (Fig. 7.5a) by BASF following Caro's initial work for the company required sulfonation of anthraquinone to provide the essential sulfonic acid. Because of this, BASF needed ever-increasing volumes of fuming sulfric acid (oleum). The key suppliers were the vitriol distilleries in Bohemia, who produced the sulfuric acid but they could not keep pace with the rising demand; oleum became scarce and expensive. Although BASF set up sulfuric acid production as early as 1866, it was Rudolf Knietsch, one of BASFs chemists (1884–1906) who developed an economical alternative process in 1888. His sulfuric acid contact process, initially using platinum on asbestos, made BASF the largest sulfuric acid producer in the world at that time. Thus, Heinrich Caro was more than aware of these needs and he experimented with sulfuric acid. However, it was not until 1898 that he published his work on the use of persulfates in the oxidation of anilines and reported on peroxysulfuric acid (H_2SO_5) [28]. This has become known as Caro's acid and comes from admixture of equimolar quantities of hydrogen peroxide and sulfuric acid:

$$H_2O_2 + H_2SO_4 \rightarrow H_2SO_5 + H_2O$$

One of Heinrich Caro's last industrial exploits was the successful production of pure hydrogen. Carl von Linde, the founder of the now multinational industrial gases and engineering company (1879) sought the collaboration of Adolf Frank and Heinrich Caro to find a way to produce pure hydrogen from water gas. The outcome was the 1909 Linde-Frank-Caro method whereby carbon dioxide and hydrogen are removed from water gas by condensation. Invented in 1909 by Adolf Frank, it was developed with Linde and Caro to become the most important method for hydrogen production. It involved water gas being compressed to 20 bar and pumped into a reactor where a water column removed most of the carbon dioxide and sulfur. Tubes with caustic soda then removed the remaining carbon dioxide, sulfur, and water from the gas stream. The gas then enters a chamber where it is cooled to -190 °C, thus condensing most of the gas to liquid. The remaining gas is pumped to the next vessel where the nitrogen is liquefied by cooling to -205 °C, leaving hydrogen gas as the product.

Heinrich Caro was also dedicated to the betterment of science workers and he co-founded the Association of German Engineers in 1856, was a member of the Society of Chemical Industry, and the Society of German Chemists (now the Deutsche Chemisches Gesellshaft) and its chairman from 1898 to 1900. He received honorary doctorates from Munich (1877), Heidelberg, TH Darmstadt and, at the Perkin 25th celebration of mauvine, Leeds (1906). His daughter Amalie put his works in trust as a Special Collection with the Deutsches Museum in Munich (The Caro Nachlass).

There can be no doubt that Heinrich Caro was responsible for more of Germany's successes in the dye industry than any other individual. He died while on holiday visiting Dresden after being in Berlin and he is buried in Mannheim. Being first in the introduction of so many new dyes and cultivating close and fruitful contacts with academic chemists, Heinrich Caro attained almost a mythical status after his death.

References

1. ChemieFruende Erkner e.V. (2013) Caro, Heinrich. http://www.chemieforum-erkner.de/chemie-ges-chichte/personen/caro_heinrich.html. Accessed 27 Mar 2015
2. Travis AS (1998) "Ambitious and Glory Hunting. Impractical and fantastic": Heinrich Caro at BASF. Technol Cult 39:105–115
3. Travis AS (1991) Heinrich Caro at Roberts, Dale & Co. Ambix 38(3):113–134 and references cited therein
4. Morris PT, Travis AS (1992) The Chemical Society of London and the Dye Industry in the 1860s. Ambix 39(3):117–126
5. Meth-Cohn O, Smith M (1994) What did W. H. Perkin actually make when he oxidised aniline to obtain mauveine? J Chem Soc Perkin 1:5–7

6. Sousa MM, Melo MJ, Parola AJ, Morris PJT, Rzepa HS, Seixas de Melo JS (2008) A study in Mauve: unveiling Perkin's Dye in historic samples. Chem Eur J 14:8507–8513
7. Travis AS (2007) Anilines: historical background. In: Rappoport Z (ed) The Chemistry of Anilines, Part 1. Wiley, Chichester, pp 1–75
8. Caro H, Wanklyn JA (1867) On the relation of Rosaniline to Rosolic Acid. Proc Royal Soc 15:210–213
9. Anon (1863) Anile blue. Chem News 7:264
10. Hofmann AW (1863) Note on the composition of aniline-blue. Proc Royal Soc 12:578–579
11. Travis AS (1992) Science's powerful companion: A. W. Hofmann's investigation of aniline red and its derivatives. Brit J Hist Sci 25:27–44
12. Griess P (1858) Vorläufige Notiz über die Einwirkung von salpetriger Säure auf Amidinitro- und Aminitrophenyl-säure. Ann Chem Pharm 106:123–125
13. Heines V (1958) Peter Griess—discoverer of diazo compounds. J Chem Educ 35:187–191
14. Kehrmann F, Klopfestein W (1923) Neue Synthesen in der Gruppe der Chinonimid-Farbstoffe, III.: Synthesen der Induline 3 B und 6 B. Ber Dtsch Chem Ges 56(11):2394–2397
15. Kehrmann F (1924) Nouvelles synthéses dans le groupe des matières colorantes dérivées des quinone-imines. Helv Chim Acta 7:471–473
16. Kehrmann F, Stanoyévitch L (1925) Nouvelles synthèses dans le groupe des matières colorantes dérivées des quinone-imines, IX. Synthèse totale de l'induline 6 B. Helv Chim Acta 8:661–663
17. Kehrmann F, Stanoyévitch L (1925) Sur le monophényl-tétraminobenzène et quelques-uns de ses dérivés. Helv Chim Acta 8:663–668
18. da Conceição Oliveira M, Dias A, Douglas P, de Melo JSS (2014) Perkin's and Caro's Mauveine in Queen Victoria's Lilac Postage Stamps: a chemical analysis. Chem Eur J 20:1808–1812 and references cited therein
19. BASF History. https://www.basf.com/global/en/who-we-are/history.html. Accessed 20 Jan 2019
20. Graebe C, Liebermann C (1868) Ueber Alizarin und Anthracen. Ber Dtsch Chem Ges 1:49–51
21. Graebe C, Liebermann C (1869) Ueber künstliche Bildung von Alizarin. Ber Dtsch Chem Ges 2:14
22. Graebe C, Liebermann C (1869) Improved process of preparing Alizarine. United States Patent 95,465
23. Caro H, Graebe C, Liebermann C (1870) Ueber Fabrikation von künstlichem Alizarin. Ber Dtsch Chem Ges 3:359–360
24. Badische Anilin- und Sodafabrik [BASF] (1877) Verfahren zur Darstellung blauer Farbstoffe aus Dimethylanilin und anderen tertiaren aromatischen Monaminen. Deutsches Reich Patent 1886. See also: Caro H (1878) Improvement in the production of dye-stuffs from methyl-aniline. United States Patent 204,796
25. Flight W, Meyer A, Michselis, Oppenheim (1873) Die Chemie auf der 46 Versammlung Deutscher Naturforscher und Aerzte zu Wiesbaden. Ber Dtsch Chem Ges 6:1387–1416, 1390–1392
26. Caro H, Graebe C (1878) Zur Kenntniss der Rosolsäuren. Ber Dtsch Chem Ges 11:1348–1351
27. Caro H (1892) Ueber die Entwickelung der Theerfarben-Industrie. Ber Dtsch Chem Ges 25:955–1105
28. Caro H (1898) Zur Kenntniss der Oxydation aromatisher Amine. Z Angew Chem 11:845–846

Chapter 8
John William Draper (1811–1882)

Anyone familiar with photochemical phenomena should be aware of the name 'Draper', as it is enshrined in the first law of photochemistry—the Grotthuss-Draper law. These two 19th century scientists recognised that for light to produce an effect upon matter it must first be absorbed. Although Theodor von Grotthuss (1785–1822) proposed the law first in 1817, he gained more recognition for his electrochemical work and his idea that charge is not transported by the movement of particles but by the breaking and reformation of bonds—the Grotthuss mechanism. This was the first essentially correct concept for charge transport in electrolytes, and is still valid for charge transport in water; the current proton hopping mechanism is a modified version of the original. However, it was Draper (Fig. 8.1) who gave substance to the proposal that light needs to be absorbed before it impacts on matter. This he independently recognized in 1842. He trained as a chemist and physician who conducted investigations and wrote on chemistry, physics, botany, physiology, photography, telegraphy, teaching, history, sociology, and religion [1, 2].

John William Draper was born on 5 May 1811 in the parish of St. Helens, near Liverpool in Lancashire, England, the son of Wesleyan clergyman Rev. John C. Draper and Sarah (Ripley) Draper. Draper senior had a moderate income but, because of his ministry, the family moved frequently from parish to parish throughout England. Until 1822 the young John was home-tutored by his father, who had an interest in scientific subjects, and by private tutors. Only at 11 years of age did he enter school—Woodhouse Grove in Leeds. This school had been founded by the Methodist Conference in 1812 to educate the sons of ministers, and John Wm. stayed there for four years until 1826, when he once again returned to home instruction. He showed an early interest and ability in science and in 1824 was selected to deliver the customary school address to the Wesleyan conference, which met in Leeds that year. In 1829 he entered the (then) recently opened University of London (it became University College in 1836) to major in chemistry under Dr. Edward Turner, author

---

A previous version of this chapter was published as Halton B (2013) John William Draper (1811-1882). Chem N Z 77:136–141.

8 John William Draper (1811–1882)

Fig. 8.1 Left: John William Draper, 1811–1882 (from *Popular Science Monthly* Volume 4 via Wikipedia), right: Dorothy Catherine Draper (by John Draper from Wikipedia)

of the first English textbooks on organic chemistry. It was here that Draper gained an interest in the chemical effects caused by light, an interest that drove his subsequent career. In 1831, during his undergraduate education, he married Antonia (Ann) Gardner, the daughter of a court physician who served the Emperor of Brazil, Dom Pedro [2], and who had been living with an aunt in London [1]. Later that same year John's father died unexpectedly and John Wm. never completed his London degree but was awarded a "certificate of honours" in chemistry instead [2]. His mother, his three sisters, and he and his new wife were persuaded to immigrate to the US in 1832 by relatives who had settled in Christiansville (now Chase City), Mecklenburg County, Virginia. Before the American Revolutionary War, several maternal relatives had migrated there and had founded a small Wesleyan colony where John Wm. hoped to gain a teaching post at the nearby Methodist College.

The family's arrival in the US proved to be too late for John Wm. to gain the prospective teaching post, so he established a laboratory in Christiansville and set to scientific experimentation on capillarity. He then published on this and a variety of other scientific subjects. He had had three papers published while in London and eight more came before he was able to further his education at the University of Pennsylvania. From 1835, he studied chemistry and physics under Dr Robert Harris while taking the university course in medicine. He also had chemistry classes from John K. Mitchell, at that time Professor of Chemistry at the Franklin Institute in Philadelphia. His sister Dorothy Catherine funded his education. In London, she had taught art and, once in Virginia, she and her sisters, Elizabeth and Sarah, opened the *Misses Draper Seminary for Girls*. Here she continued to teach drawing and painting and by 1835 had saved sufficient money to fund her brother's education [3]. John Wm. Draper graduated in medicine in March 1836 and was soon employed to teach at Hampden Sydney College in Virginia. This school, established to form good men and good citizens in an atmosphere of sound learning, had learned of Draper's abilities.

However, his teachers motivated him to accept an appointment the following year at New York University, where he became Professor of Chemistry and Botany in 1838. By then he was the father of two boys. He was one of the founders of the New York University Medical School becoming a professor in the school of medicine from 1840, and its president and professor of chemistry from 1850 until 1881. At the time, he joined the Medical School there were few students, but with his enthusiasm and the help of his friend J. G. Bennett, editor and publisher of the *New York Herald*, the 1841–1842 session had some 240 students. He raised two further boys and two girls in New York.

Leaving his historical writings aside (see below), John Draper's scientific work would span chemistry, physics, medicine, botany and scientific literature. From his undergraduate days in London, there were papers on volcanoes and the formation of the Dead Sea, and one on electrical decompositions was published in 1833. However, it was from 1834 that his independent studies appeared. The first, a letter to the editor of the *American Journal of Science* [4], appeared in 1834 and was one of a series of studies on capillary action that led to definitive results on osmosis and on the circulation of sap in plants and blood in animals. The first paper concerning the action of light was the following year, when he reported the results of a study on whether light exerts a magnetic action. His first purely chemical paper was the "Chemical Analysis of the Native Chloride of Carbon, a singular mineral" published in the *Journal of the Franklin Institute* (of the State of Pennsylvania) in 1834 [5]. The "native chloride of carbon" had been collected on the Isle of Sheppey (off the Kent coast and in the Thames estuary, England) in the summer of 1832 and, although most had been given to Prof Turner at London University, he had taken some 12 g (200 grains) of the sample to Christainsville. His studies concluded that the compound contained two atoms of carbon and one of chlorine and that the presence of hydrogen seemed unlikely. His concluding statement was:

> The production of this substance, to judge by its scent at first, and the locality in which it was found, seems to be referable to a marine animal. But through what singular changes must a dead fish pass, before its remains would leave a chloride of carbon, nearly pure?

With the exception of well-recognized marine products such as Tyrian purple [6], this, though incorrect, is one of the earliest chemical analyses of a marine species.

This paper was followed the next year by one on the analysis of coins and medals [7] and, in 1836, a practical set of instructions and commentary for carrying out "microscopic chemistry" [8]. Draper's premise here was that much good chemistry could be carried out using semi-micro techniques (as we would now call them). It is particularly interesting to see him advocating the use of sand and water baths (admittedly in a domestic frying pan and domestic pan, respectively) for heating purposes, holding a hot tube by its neck with folded paper, and the use of apothecary phials for small scale distillation and sublimation. It needs to be remembered that in the 1830s organic chemistry was in its infancy, that only 55 elements were recognised (in 1835,) and the few instruments available limited the measurement of chemical and physical properties. Many of the traditional early studies had been conducted using large retorts and big furnaces in expensive laboratories. Draper's premise was that chemistry should be available and to able to be performed by anyone with an aptitude and skill for it, and his microscopic chemistry fitted well to this.

John Draper's career as a research scientist flowered from 1839 to 1856 as Professor of Chemistry at New York University, where he published some 45 papers and five textbooks, although his writing continued well into the 1870s. He published two chemistry books and others on physiology, natural philosophy and botany. His dominant contribution was on the constitution of "radiant energy" which led to some 28 publications. Some of his earliest investigations were directed to determining the various forms of energy that exist in solar light. In common with his contemporaries, he regarded heat and light as imponderable agents, entirely different from one another but coexisting in solar light. However, by the early 1870s, he had accepted the essential unity of radiant energy, and subsequently was adept in discussing the chemical force in the spectrum [9].

Draper approached radiant energy by studying the chemical action of light to determine the effect of the different kinds of light in photography, on chlorine gas, and on the growth of plants. Most studies of light from 1725 had involved its action on silver salts, leading to Scheele distinguishing radiant heat and light in 1777, and then the concept of "chemical rays" from effects in the violet region of the spectrum by Ritter and Wollaston in the first years of the 19th century [2]. For many years, Draper had studied the various forms of energy that exist in solar light, with the chemical action of light being a particular fascination. He performed many experiments to distinguish the effects of the different forms of light on chemical change. He studied the effect of light in changing the colours of metallic salts and was quite accustomed to applying the photographic process in the solution of physical problems prior to Louis Daguerre's 1839 discovery. This was a photographic process which became known as the daguerreotype and involved a direct positive being made in the camera on an iodine-treated silvered copper plate. Draper gained information of the process from Samuel Morse, at that time Professor of Fine Arts at the university and a close friend (see below) who had been in Paris. Draper improved the process by exposing the plates to bromine fumes, finding that the mixture of silver iodide and bromide was much more sensitive to light. This was to the extent that in the same year (1839), he took photographs of the human face, being the first to use the daguerreotype in New York. He had the process so perfected [10] by early 1840 that the image he took of his sister Dorothy Catherine (Fig. 8.1) is the oldest surviving human portrait photograph. It has often been misconstrued as the first human portrait [3]. It is undoubtedly one of the earliest portraits and certainly the first of quality; it was taken with a 65-s exposure. That same year he presented the Lyceum of Natural History of New York with the first photograph of the moon's surface and improved it over the ensuing years to show craters. The original came from a 20-min exposure through a telescope giving an image about 2.5 cm in diameter. Draper perfected the process of portraiture enough [10] to instruct in the art and, with Morse, opened the first ever portrait studio on the roof of his house on Fourth Street in New York in 1840.

8 John William Draper (1811–1882)

(a) Draper: $\quad Cl_2 \xrightarrow{h\nu} 2Cl\bullet \ ; \ Cl\bullet + H_2 \longrightarrow HCl + H\bullet$

Gomberg: $\quad Ph_3CH + Zn \longrightarrow Ph_3C\bullet + ZnCl$

(b) $6CO_2 + 6H_2O \xrightarrow{h\nu} C_6H_{12}O_6 + 6O_2$

Fig. 8.2 a The first documented laboratory photoreaction, Draper's photodecomposition of chlorine, and the formation of the trihenyl methyl radical by Gomberg; **b** photosynthesis and oxygen production

The effect of light on chlorine gas provided what appears to have been the first documented example of a photochemical reaction in a laboratory [11].[1] In 1843, Draper announced to the British Association that chlorine underwent a significant change under the influence of sunlight. He described this as occurring by the absorption of "chemical rays" which changed the character of the gas to the extent that it would then unite with hydrogen gas, a property not seen from chlorine kept in the dark or with hydrogen. While we now find it easy to accept that the chlorine molecule absorbs a photon, dissociates into a pair of chlorine atoms and abstracts a hydrogen atom to produce HCl with the propagation of a chain reaction (Fig. 8.2a), none of this was a part of 1840 thinking—the first persistent radical was the triphenylmethyl radical of Gomberg in 1900.

From this study, Draper concluded that solar light consisted of a third and new imponderable: the agent that produced chemical change. He gave this the name "tithonicity" from Roman mythology. Tithonus was a beautiful youth with whom Aurora fell in love and married in heaven. Although he became immortal, he became feeble and decrepit—unlike his bride, lost all his strength and had to be rocked to sleep. Likewise, solar light was thought by Draper to be weakened from its interaction with matter. In addition to "tithonicity", Draper extended his usage to include "tithonography", "tithonic effect", "tithonometer" and "tithonoscope" as "music in an English ear" [11] (See Footnote 1). That year he used this property (the decomposition of chlorine and formation of HCl) in the design of an actinometer to measure the force of the tithonic rays—his tithonometer [12]. The Draper actinometer was subsequently improved upon by Bunsen and Roscoe [2] who refined it to the extent that it allowed not only accurate comparative determinations, but reduced the chemical action of light to an absolute measurement [13].

As early as 1837, Draper had studied the impact of solar light on the decomposition of leaves, but his early study had involved plants under coloured glass and was inconclusive. Subsequently, he germinated the seeds under glasses of different colour. Under red and violet glass, the plants behaved as in the dark, while under yellow glass they promptly assumed a green colour and behaved normally. The experiment was repeated by germinating the seeds in the dark and then exposing the young plants to

---

[1]Barker GF (1886) Memoir of John William Draper, 1811–1882. National Academy of Sciences, Washington, pp. 364–365.

the differently filtered light to assess its impact; the plants under the yellow glass grew readily, but those under red and violet not so [14]. He then assessed the decomposition of carbonic acid by grass under different light. Thus, water was degassed and saturated with the CO_2, and to it was added grass whose surfaces had been carefully freed of air. The sample was divided among seven tubes and the effect of the seven colours of the spectrum examined. The tubes onto which orange, green and yellow rays fell began to evolve gas within a few minutes and the volumes given off measured over a given time period as: Yellow-Green: 43.00 volumes; Red-Orange: 24.75 volumes; Green-Blue: 4.10 volumes. The other colours gave no gas evolution and were without effect—neither photosynthesis nor the generation of oxygen (Fig. 8.2b). Draper read his paper giving these results to the centenary meeting of the American Philosophical Society, publishing it with the full details [15, 16].

Despite these early studies, it was his spectral investigations that led to Draper's understanding of radiant energy. Thus, in 1842 he proposed that only light rays that are absorbed can produce chemical change [17]. This, now known as the Grotthuss-Draper law, is the first law of photochemistry as noted above. Draper's name is linked with that of Grotthuss who had reached the same conclusion some 25 years earlier, but had not been recognised for it. Furthermore, Draper subsequently concluded that "every part of the spectrum, visible and invisible, can produce chemical change and can modify the molecular arrangement of bodies", and that "the rays effective in producing chemical or molecular changes in any special substance are determined solely by the absorptive power of that substance".[2] The studies prior to 1843 had used prismatic refraction, but that year he persuaded Joseph Saxton of the US Mint in Philadelphia to rule a diffraction grating for him. The results obtained from silvering the surface with tin amalgam, provided a far more intense reflected spectrum than the former transmitted ones. Draper suggested that the best result would come from ruling the grating on steel or speculum metal (a mixture of approx. two-thirds copper and one-third tin that makes a white brittle alloy capable of being polished to a highly reflective surface). He also suggested that the regions of the spectrum should be denoted by their wavelengths, saying: "The measures of one author will compare with those of another and the different phenomena of chemical changes occurring through the agency of light become allied at once with a multitude of other optical results".[3] By using the grating, the first photograph of the solar spectrum was obtained. Draper used a daguerreotype plate sensitised by iodine and then bromine. This was exposed for 30 min giving maximum sensitivity at wavelength 0.00001538 Paris inch.[4]

In 1847, he published the observation that all solids glow red at about the same temperature (798 K), and this is now known as the Draper point [18], the approximate temperature above which almost all solid materials visibly glow as a result of black-

---

[2]Barker GF (1886) Memoir of John William Draper, 1811–1882. National Academy of Sciences, Washington, pp. 366.

[3]Barker GF (1886) Memoir of John William Draper, 1811–1882. National Academy of Sciences, Washington, pp. 368.

[4]A Paris inch, or pouce, is an archaic unit of length, then common for giving the measurement of lenses. It could be subdivided into 12 ligne (Paris line), and 12 Paris inches made a Paris foot. The lens measurement of Fraunhofer 9-inch telescope, the largest of its day, was Paris inches (24 cm).

body radiation. In addition, he concluded that as the temperature of an incandescent body is increased it emits rays of light of an increasing refrangibility. At that time the science of spectroscopy had evolved to the extent that all glowing solids were known to emit continuous spectra, unlike gases that emitted bands or lines. Draper went beyond defining what has become known as the Draper point to study the spectral composition of flames (the production of light by chemical action [19]) using a slit, a prism and a telescope. By examining a range of compounds (that included alcohol and its solutions with boric acid and strontium nitrate, oil, phosphorus, sulfur, carbonic oxide (CO), hydrogen, cyanogen (NC-CN) and H_2S) he concluded that, notwithstanding the diversity in colour, all the flames yield the same results. Every prismatic colour was found in them, even bright Fraunhofer lines of different colour, and even in those cases where the flame is very faint. The spectrum of cyanogen was, he said, "so beautiful that it is impossible to describe it in words or depict it in colours". Subsequently he came to the conclusion that "the occurrence of lines, whether bright or dark, is hence connected with the chemical nature of the substance producing the flame and that if we are ever able to acquire certain knowledge of the physical state of sun and other stars, it will be by an examination of the light they emit" [20]. In 1879, Draper built a spectrometer based upon the photometric method of Bouguer that he had used successfully in 1847. However, this instrument measured light intensity. He found that the colours of a gas flame and sunlight both disappeared in the reverse order of their refrangibility, red being the last to disappear. When the shutter admitting the daylight was slowly opened, it was the extreme violet that disappeared first, red last; and on closing the shutter it was the red that appeared first. On reducing the extraneous light intensity all came into view at the same time.[5] These experiments proved that the apparent brightness of yellow, as seen by the eye, then accepted as the most intense in the spectrum, is a purely physiological phenomenon.

Another area studied by Draper was phosphorescence [21]. His first report [22] included a general discussion of the phenomenon starting from belief in the existence of the carbuncle, a stone supposed in the very infancy of chemistry to have the property of shining in the dark. He went on to define the phenomenon to include those bodies that shine in the dark after exposure to light or on heating, and then described his experiments with John William on the fluorspar variety known as chlorophane. He selected this material because it could be obtained as both almost opaque and perfectly transparent, was easily cut and polished into almost any shape, and yields a superb emerald-green light. He found that when phosphorescence was triggered by an electric spark there was no perceptible change in volume or any molecular change detectable by polarized light. However, there was a small amount of heat generated. He concluded that the quantity of light emitted by a phosphorescent body was proportional to the intensity of the light to which it had been exposed. He went on to show that the rays which generate phosphorescence were the violet rays.

---

[5]Barker GF (1886) Memoir of John William Draper, 1811–1882. National Academy of Sciences, Washington, pp. 369–370.

As early as 1834 Draper had studied the action of the galvanic battery and published design and construction improvements, and went on to describe the construction of a torsion galvanometer in 1839. However, it is his 1843 paper on the law of the conducting power of wires that deserves mention [23]. This was a study he performed to help Samuel Morse in perfecting his single wire telegraph, and showed that the diminution in strength of an electric current reduces rapidly with increasing wire length and that the conducting effect is represented by a logarithmic curve. Although Morse was not the first to perfect the telegraph, his code for word transmission—the Morse code—was universally adopted.

It is clear that John William Draper was a pioneer in photochemistry and a spectroscopist of exceptional ability and acumen. He had no doctorate in science but was a 19th century physician gone astray. Yet he was popularly known for other things. He had many papers in the medicinal chemistry area and, in his later life, he gained fame as an author on account of his published books: *History of the intellectual development of Europe* (1862) [24], *Thoughts on the future civil policy of America* (1865) [25], and *A text-book on physiology* (1866) [26]. His noted *History of the American Civil War* (for which he was given access to whatever records he needed by the then US Secretary of War) ran to three volumes, published [27] between 1867 and 1870, and his most well-known *History of the conflict between science and religion* [28] was published in 1874. This last treatise went through more than twenty editions in its first ten years and was translated into most European languages. Because Draper was open and honest about the continuing conflict between science and religion, his book was placed on the Index Expurgatorius of the Catholic Church.

There is one further role in Draper's life that few are aware. He became inaugural President of the American Chemical Society [29]. In 1874 a group of about 70 chemists had met in Northumberland, Pennsylvania on 1 August to celebrate the centenary of the death of Joseph Priestley and, while some subsequently joined the American Association for the Advancement of Science, it became clear that the discipline of chemistry needed its own organisation. Charles F. Chandler, Professor of Chemistry in the School of Mines at Columbia University, was the main driving force for this, and on 6 April 1876, a group of 35 chemists met in the University of New York. Among those present was William Henry Nichols, a 24-year old chemical entrepreneur and former student of Draper who, speaking last, argued so forcefully for the establishment of the American Chemical Society that the passage of the vote was assured. The ACS came into existence that evening, a constitution was adopted and the New York participants selected officers. Chandler was an obvious choice as president because he was well known in New York, but he believed that the new organisation needed to be a national one that needed a president whose fame and reputation were known to the general public, who would attract members nationally, and who would have the society accepted by the organisations in Europe. In Chandler's mind there was only one person who met these criteria, and that was John Wm. Draper. At 65 years of age, Draper was invited and accepted the position on 11 April. He advised, however, that his health would prevent his active participation in the society, as he suffered from severe attacks of urinary stones, which prevented him from travelling [2]. Thus, Draper's name and reputation were used more than his active participation. He never chaired a monthly meeting and was unable to attend a dinner for foreign chemists attending the Philadelphia Centennial Exhibition in June that year. However, he did give his inaugural Presidential Address in November in New York City under the title Science in America [30]. Draper died at his New York home in Hastings on 4 January 1882.

Among the other awards accorded John Draper were membership of many learned societies in the US and Europe. He gained the Rumford medals of the American Academy of Arts and Sciences (1875), the founding presidency of the American Union Academy of Literature, Science and Art (1869), Chairman of the American Photographic Association (1864) and an LLD from the college of New Jersey at Princeton. His house in Hastings-on-Hudson stands in Draper Park, now an historic nine-acre property containing five nineteenth-century structures in a relatively intact landscape setting [31]. The entire site, buildings and park, was designated National Historic Landmark in 1975.

References

1. Barker GF (1886) Memoir of John William Draper, 1811–1882. National Academy of Sciences, Washington
2. Wisniak J (2013) John William Draper. Educ quim 24(2):214–223
3. McManus HR (1995) The most Famous Daguerreian Portrait: exploring the history of the Dorothy Catherine Draper Daguerreotype. In: Baty LA (ed) The Daguerreian annual 1995. The Daguerreian Society, Pittsburgh, pp 148–171
4. Draper JW (1834) Influence of electricity on capillary attraction. Am J Sci Arts 26:399–400
5. Draper JW (1834) Chemical analysis of the native Chloride of Carbon, a singular mineral. J Franklin Inst 18:295–298
6. Ziderman II (1986) Purple Dyes Made from Shellfish in Antiquity. Rev Prog Color Relat Top 16:46–52
7. Draper JW (1835) Coins and medals. Am J Sci Arts 29:157–166
8. Draper JW (1836) Observations on microscopic chemistry. J Franklin Inst 22:378–384
9. Draper JW (1873) Researches in actino-chemistry. Memoir Second. On the Distribution of Chemical Force in the Spectrum. Am J Sci Arts 5:25–38, 91–98
10. Draper JW (1840) On the process of Dguerreotype, and its application to taking Portraits from the life. Philos Mag 17(3):217–225
11. Draper JW (1843) On a change produced by exposure to the beams of the sun in the properties of an elementary substance. Rep Br Assoc Sci Pt. 2:9
12. Draper JW (1843) Description of the Tithonometer, an instrument for measuring the chemical force of the indigo-tithonic rays. Philos Mag 23:401–415
13. Bunsen RE, Roscoe HE (1847) Photo-chemical researches. Part I. Measurement of the chemical action of light. Philos Trans R Soc Lond 147:355–380
14. Draper JW (1837) Experiments on solar light. J Franklin Inst 19: 469–479, 20: 38–46, 114–125, 250–253
15. Draper JW (1843) On the decomposition of carbonic acid and the alkaline carbonates by the light of the sun. Proc Am Philos Soc 3:111–114
16. Draper JW (1843) On the rapid detithonizing power of certain gases and vapours, and on an instantaneous means of producing Spetral appearances. Philos Mag 22:161–165
17. Hentschel K (2002) Why not one more imponderable? John William Draper and his Tithionic rays. Found Chem 4:5–59

18. Draper JW (1847) On the production of light by heat. Philos Mag 30:345–360
19. Draper JW (1848) On the production of light by chemical means. Philos Mag 32:100–114
20. Draper JW (1857) On the diffraction spectrum. Remarks on M. Eisenlohr's recent experiments. Philos Mag 13:153–156
21. See e.g., Coles MP (2013) Fluorescent minerals. Chem N Z 77:42–47
22. Draper JW (1851) On the phosphorescence of bodies. Philos Mag 1:81–100
23. Draper JW (1843) On the law of the conducting power of wires. Am J Sci Arts 45:392–394
24. Draper JW (1862) History of the intellectual development of Europe. Harper, New York
25. Draper JW (1863) Thoughts on the future civil policy of America. Harper, New York
26. Draper JW (1866) A textbook on physiology. Harper, New York
27. Draper JW (1867–1870) History of the American Civil War, vol. 3. Harper: New York, Harper
28. Draper JW (1874) History of the conflict between science and religion. Appleton, New York
29. John W (1876) Draper and the founding of the American Chemical Society. ACS National Historic Chemical Landmark. https://www.acs.org/content/acs/en/education/whatischemistry/landmarks/draperacs.html. Accessed 24 Jan 2019
30. Draper JW (1876) Science in America. Inaugural address of Dr. John W. Draper, as president of the American Chemical Society. John F. Trow & Son, New York
31. Images of the property that include the observatory can be seen on the Stephen Tilly (architect) website: http://www.stillyarchitect.com/portfolio/preservation/draper/draper.htm. Accessed 20 May 2013

Chapter 9
Sir Thomas Hill Easterfield (1866–1949)

Thomas Hill Easterfield (Fig. 9.1) was born on March 4, 1866 in Doncaster, Yorkshire, England, the son of Edward and Susan (neé Hill). His father was a banker who rose to become secretary and then manager of the local branch of the Yorkshire Savings Bank. Thomas Easterfield entered Doncaster Grammar School, winning prizes in Classics, divinity, French, German and science during his time there [1, 2]. His younger years had him contemplating a career in the textile industry whilst studying at Yorkshire College Leeds (the forerunner of Leeds University). In May of 1881 he won one of the Akroyd Scholarships[1] (in geology [3]) valued at £20 annually for two years to attend the college. He subsequently became a Brown Scholar and published two papers, one on photography [4] and the other on a glacial deposit [5]. His time at Yorkshire College led to the award of a B.A. degree [1–3, 6–9], according to his October 29, 1931 invited founding member application to join the New Zealand Institute of Chemistry (NZIC) (Fig. 9.2).[2] Initially he was an Associate, then ten years later in 1941 he became a Fellow. From Leeds, Thomas proceeded to Cambridge's second oldest college, Clare, as a senior foundation scholar studying geology, physics and chemistry, and was awarded a first-class degree with honours in chemistry and geology in the natural science tripos in 1886. Apart from his academic acumen Easterfield was a notable middle-distance runner, representing the University in the mile and three-mile events.

Following his time at Cambridge, Easterfield gained postgraduate experience at the Zurich Polytechnic College and the University of Zurich in Switzerland, and then at Würzburg University on the Main river in Germany. There he worked under the noted organic chemist Emil Fischer who had him study two topics, one relating to the structure of sugars and the other to the three-dimensional structure of carbon

---

[1] Data reported in the Leeds Mercury, 14 May 1881.
[2] I thank Richard Rendle for providing the application from the Institute archive.

A previous version of this chapter was published as Halton B (2017) Sir Thomas Hill Easterfield (1866–1949). Chem N Z 81:83–88.

© Springer Nature Switzerland AG 2020
B. Halton, *Some Forgotten Chemists*, Perspectives on the History of Chemistry,
https://doi.org/10.1007/978-3-030-16403-4_9

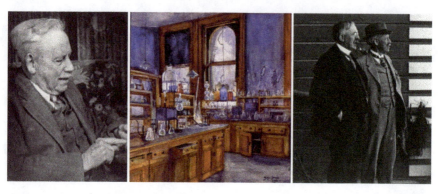

Fig. 9.1 Left: Thomas Hill Easterfield (VUW-2-163), centre: Sybil Johnson: Victoria College: First Chemistry Laboratory, 1901. 1901 watercolour on paper. Victoia University of Wellington (VUW) Art Collection, gift of Lady Easterfield; right: Sir Ernest Rutherford and T. H. Easterfield, Nelson, 1925 (from VUW-S19). (each with permission from VUW, New Zealand)

Fig. 9.2 T. H. Easterfield's 1931 application for admission to the NZIC

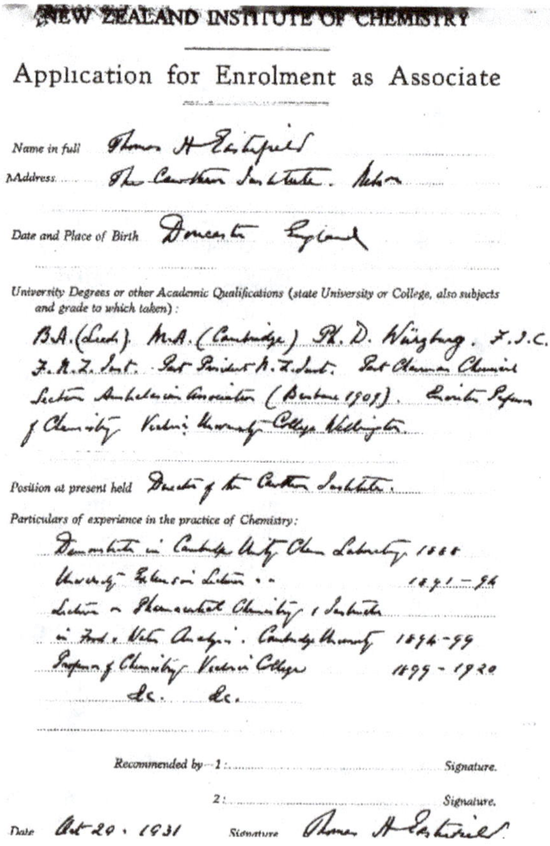

NEW ZEALAND INSTITUTE OF CHEMISTRY

Application for Enrolment as Associate

Name in full Thomas H Easterfield

Address The Cawthron Institute. Nelson

Date and Place of Birth Doncaster England

University Degrees or other Academic Qualifications (state University or College, also subjects and grade to which taken):

B.A. (Leeds). M.A. (Cambridge) Ph. D. Würzburg. F.I.C. F.N.Z. Inst. Past President N.Z. Inst. Past Chairman Chemical Section Australasian Association (Brisbane 1909). Emeritus Professor of Chemistry Victoria University College Wellington.

Position at present held Director of the Cawthron Institute.

Particulars of experience in the practice of Chemistry:

Demonstrator in Cambridge Univy Chem Laboratory, 1888
University Extension Lecturer .. 1891–94
Lecturer on Pharmaceutical Chemistry & Instructor in Food & Water Analysis. Cambridge University 1894–99
Professor of Chemistry Victoria College 1899–1920
&c. &c.

Recommended by—1: .... *Signature.*

2: .... *Signature.*

Date Oct 29. 1931 *Signature* Thomas H Easterfield

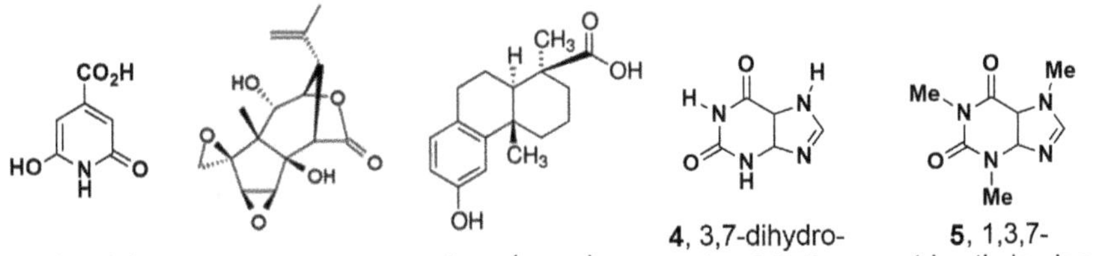

1, citrazinic acid; **2**, tutin; **3**, podocarpic acid; **4**, 3,7-dihydro-purine-2,6-dione (xanthine); **5**, 1,3,7-trimethylpurine-2,6-ione (caffeine)

Fig. 9.3 Some of the molecules Easterfield prepared

compounds [1]. His Ph.D. work on the chemistry of citrazinic acid (**1**) (Fig. 9.3) was also started there. His doctoral degree was awarded in 1894 based on a 35-page thesis entitled: *Zur Kenntnis der Citrazinsäure* (knowledge of citrazinic acid).

There is some confusion as to precisely when Easterfield went to Würzburg, but he returned to Cambridge in 1888 as a lecturer under the University Extension Movement and worked in the organic chemistry laboratory according to Marsden [7], or as Davis [1] gives it a demonstrator in the chemical laboratory, became university extension lecturer and was, for a while, part-time science master at Perse Grammar School in Cambridge. In contrast, MacFarlane [2] tells us that Easterfield appears to have returned to England several times during his Ph.D. studies and gained a variety of practical experiences, especially over the 1891–1894 period and prior to the award of his Ph.D. degree. Irrespective of the precise detail here, Easterfield's publications on the topic of his Ph.D. are jointly with W. J. Sell [10–12], a demonstrator at Cambridge who had published from that position On the Volumetric Determination of Chromium in the *Transaction of the Chemical Society* in 1879. Moreover, Sell published on derivatives of citric and aconitic acids with Easterfield in 1892 and continued with the citrazinic acid study until after Easterfield left Cambridge, publishing the results with F. W. Dootson in 1897 and with H. Jackson in 1899. It is the view of this author that much, if not most, of the Easterfield Ph.D. experimental work could have been performed in Cambridge with Sell as the assistant supervisor. Easterfield's ability to teach, lecture, and research grew over the 1887–1894 period as judged from his twelve publications from that period.[3] As important is the fact that during his time in Germany, Thomas met and courted Bavarian Anna Maria Kunigunda Büchel and married her in Würzburg on September 1, 1894.

From Würzburg, Easterfield returned to Cambridge in 1894 to a lectureship in pharmaceutical science and in the chemistry of sanitary science [6]. He filled these positions with success and continued his research studies publishing a further four papers on aspects of charcoal, Indian hemp resin (*Cannabis sativa*), and cannabinol. When New Zealand's Victoria College of Wellington Board chose to appoint its inaugural professors, four chairs were advertised, one being in chemistry and physics. Easterfield applied for this, was interviewed in England, and accepted the offer of appointment as announced in the New Zealand Herald, on January 13, 1899. His

[3]See Appendix entries in Ref. 2.

appointment came with funds to purchase and transport needed scientific equipment. £100 for the physics and £50 for chemistry were initially allocated, spent and the supplies dispatched, only to be damaged in transit. Subsequently, the situation was made passable with the gift of £25 from George W. Wilton, the founder of what became one of New Zealand's major importers of scientific equipment through the 20th century.

Easterfield, his wife and two daughters, joined John Rankine Brown (Classics) and his family on the steamship *Kaikoura* in Plymouth for their travel to New Zea-land on the evening of February 11th, 1899, in the foulest of weather. Hugh Mac-Kenzie (English Language and Literature) and his family had boarded in London. The Brown and Easterfield families were taken out to the ship by tender and were in a bedraggled and collapsed condition when they reached it in Plymouth Sound. When the Kaikoura put out to sea things were even worse, for she headed into a first-class gale, which lasted three days and destroyed a large part of the captain's bridge. However, the sea was calm and pleasant by the time they reached Tenerife, and on rounding the Cape the three professors had learned something of one another's idiosyncrasies [13]. The fourth appointee, the unmarried Richard Maclaurin, came independently [14].

The *Kaikoura* sailed into Wellington harbour on Saturday, April 1, 1899, one of Wellington's glorious autumnal days. If, on that 1st of April when the Council met them (the professors), they had felt a little foolish they might be forgiven. Somehow, before they left England, they had been given to understand that their college was not only adequately endowed, but actually physically in existence—that their task as founders was, as it were, to walk in and begin lecturing [14]. The reality was very different. There was no college edifice and little finance. On being met by the college councillors, Easterfield was told by one that he had gained his appointment by just one vote—and that because there was no Scotsman available! The professors were given a little time to accustom themselves to their new surroundings before a formal welcome in the offices of the Education Board on Wednesday April 12. However, the first students registered just one week after their arrival and were met by the professors. Term 1 started on April 18 with 30 students in chemistry and 11 in physics. The first two of the inaugural lectures were presented that same evening [15, 16]. In his address, *Research as a Prime Factor in Scientific Education* [17], Easterfield espoused his views on education, and to a colonial audience 118 years ago he must have seemed a revolutionary. He was the inaugural appointee of the four who came German-trained with a definite purpose. He was plain in saying that early specialisation by students with research work and original investigation should come before a degree was awarded and that a good laboratory was an absolute necessity for staff and student research in science. That some of the research could benefit the country also had to be recognised. Research was in marked contrast to the reality of Wellington College. There was no building and no facilities for science. Consideration had been given to using a boarding house on Tinakori Road (now the Prime Minister's formal residence) for the college but that was discarded after Easterfield announced that the ground floor could be suitable for science but not the upstairs bedrooms. After the term started, three upper rooms in the Technical School

Building on Victoria Street were made available [16]. It was there that Easterfield himself fashioned a laboratory out of boards and trestles.

The balance that Easterfield had been given by his Cambridge colleagues, and a prize possession, was placed on a packing case in one corner. It was here that science was taught and where Thomas Easterfield began his New Zealand researches carrying out his own experiments and directing those of his students. There was no assistant and no lab attendant (a technician in the modern idiom), and the students had to bring up the water and empty the slop buckets themselves. On one occasion this was missed and the corrosive liquid ate through the bucket, permeated the ceiling, and made a significant mess in the Technical School Director's office directly below [15, 18]. The adjacent room was the physics laboratory and lecture room, from which the students were required to remove their chairs and collapsible tables when they left—much to the annoyance of those in the room below. Yet, such primitive surroundings did not deter the research efforts. Early students gained notable college honours and the enthusiasm of the foundation professor was undeterred in establishing his discipline.

During that first year of 1899, Easterfield and Victoria College secured £3000 for equipping the laboratories. When the students arrived back for the 1900 teaching year the upstairs rooms of the Technical College had been transformed and a chemistry laboratory created, this to the extent that the arts students from below asked to come and draw it. One of these budding artists, Sybil Johnson, painted what was to Easterfield the most beautiful representation of his creation [15] (Fig. 9.1). This painting became part of his collection, subsequently to be donated to the University College by his widow. His own researches were dominated by New Zealand natural products, involving native plants, to which he had referred in his inaugural lecture. Easterfield used the laboratory fund wisely, and when the university opened its building in Kelburn in 1906 many facilities had been paid for from the residue of the fund, and with the kauri laboratory benches transferred from their city location. When tenders were called for what became the Hunter Building, Easterfield was not backward in coming forward. In his view, the instructions given to the architects were rather vague and three of the firms approached him for his opinion on the College Building. This he offered, not just to those three but to all the contenders; the edifice that stands carries many of his ideas. During his first decade at Victoria he held the chairs of chemistry and physics and was relieved of the latter only in 1909 with the appointment of Prof. T. H. Laby.

In his second year, he not only published his own research results with B. C. Aston, a local government chemist in the Department of Agriculture, but also read the paper of one of his first students, P. W. Robertson, to the New Zealand Institute (Wellington), the forerunner of the Royal Society of New Zealand. Easterfield was quick to introduce himself to the few chemists in Wellington, notably William Skey, who was the Colonial Analyst in Wellington, and Aston. With Aston, he set to work on the Tutu plant (*Coriaria*) and isolated the poisonous principal, which they named tutin (**2**) (Fig. 9.2). Their work was published on the three species of *Coriaria* found in New Zealand, namely *C. ruscifolia*, *C. thymifolia*, and *C. angustissinza*, firstly in New Zealand [19] and then in the UK in 1900 [20] and 1901 [21], respectively. Their collaboration continued with studies of karaka (*Corynocarpus laevigatus*) and rimu (*Dacrydium cupressinum*) resin, which is now known to contain podocarpic acid (**3**). Easterfield continued his work teaching, carrying out research, and directing his students, most of whom went on to notable careers in New Zealand science. He was also consultant to industry but the difficulties in funding research and the limited recognition his efforts drew began to weigh more heavily on him and his interests moved more towards agricultural chemistry from about 1908. Nonetheless, his research was recognised in 1913 with the award of the second of the New Zealand Institute's Hector Memorial Medal and Prize. Then, during the Great War with shortage of synthetic medicines he processed the opium seized by the police under New Zealand's anti-drug laws in his Victoria laboratory to provide morphine [22]. His consultancies are nicely illustrated by some of his last experiments as College Professor that were performed at the Miranui flax mill near Levin. These were to determine if industrial alcohol (ethanol) could be produced by fermenting the juices of the flax leaf. This factory was the largest and most well equipped mill of the era and his test runs produced 198.4 gallons of 95% ethanol daily. This was not used but allowed to flow down the Miranui drains! The output equates to 50,000 gallons per annum and would have been worth £3750 per year at 1s 6d per gallon in 1919 [15].

Following the death of Thomas Cawthron in 1915, the will that the gentleman made in London some thirteen years earlier was granted probate by the Supreme Court in Wellington on November 12, 1915. The bulk of the £250,000 estate was set for the erection and maintenance of an industrial and technical school institute and museum in Nelson, New Zealand, to be called the Cawthron Institute [23]. The suggestion that most of this wealthy bachelor's estate could be put to such usage was made by Cawthron's friend Joseph H. Cock when asked how best his wealth could be divided [23, 24]. As the executors had little knowledge or understanding of science they established an advisory commission to which Easterfield was appointed along with Professors Benham (Biology, Otago) and Worley (Chemistry, Auckland), the noted biologist Leonard Cockayne, and Sir James Wilson, President of the Board of Agriculture; Easterfield was appointed secretary. As with everything he did, Thomas Easterfield was thorough, conscientious and efficient. Within three months of arriving in Nelson the advisory commission's report had been sent to the trustees and gone for legal approval. As secretary, Easterfield was asked to report on the form and functions of the yet to be established institution and in 1917 he presented the aims and ideals of the trustees at a public lecture in Nelson. It became the first annual Cawthron Memorial Lecture. When it came to the appointment of the first director, the trust board had two candidates in mind—Easterfield and his former student Theodore Rigg. The older Easterfield had all the attributes that the trustees considered essential and he was appointed Founding Director in October 1919. By then he had given all that he could to establishing chemistry at Victoria University College as it then was, and he accepted the offer. He then had to live up to his comments in the Cawthron Lecture—that the Institute would have a bright future! During his 21 years at Victoria, Easterfield published no less than 26 papers [2], several of them in New Zealand, then with the Chemical Society (UK) Easterfield's research record at Victoria (University) College was solid and particularly significant as research was not high on the agenda of the New Zealand University Colleges during his tenure. He was appointed Victoria's first emeritus professor.

As founding director of The Cawthron Institute in Nelson, Easterfield found himself in a position akin to that on his arrival in Wellington. Although the Cawthron Trustees had purchased land and property on a 10.4-acre site in Annesbrook in 1917, some five kilometres from Nelson city, conversion to the facilities of a scientific Institute did not happen. Half the property was kept as an orchard almost until 1959. As in Wellington in 1899, Thomas Easterfield had to make do with temporary accommodation. This was until Fellworth House, the John Sharp property on Milton Street, was bought later in 1920 [23]. The house stood in three acres of ground and consisted of some fifteen rooms that were modified for science usage. The trustees had recognised early on that the strengths of the Cawthron would be in agriculture and horticulture, and it was in this area that Easterfield directed his attention. This contrasts with his 1917 lecture in which he was clear that research should not be restricted to applied areas but include 'pure science'.

During his first year, Easterfield appointed Theodore Rigg as his assistant and agricultural chemist and then W. C. Davies (museum curator and photographer), Drs. R. J. Tillyard (entomologist) and Kathleen Curtis (mycologist who became the first woman elected to RSNZ Fellowship in 1936), H. Harrison (librarian), and two technicians, A Philpott and E. J. Champtaloup. And then he gave each of the scientists the ability to appoint their own technicians. This engendered a good staff with excellent working relationships and loyalty to the Institute [23]. These appointments doubled the country's agricultural research capacity and were followed in 1921 by a fifth position in orchard chemistry part funded by the fruit growers association. They gave Nelson an intellectual force [24].

From taking up his position, the Cawthron Institute was the sole organisation devoted to agricultural and horticultural problems not simply in New Zealand but in the entire southern hemisphere. Requests for assistance had started to arrive shortly after Cawthron's will was made public, but Easterfield elected to address those of the Nelson region first. The Cawthron's main departments of scientific research, horticulture and agriculture were established in 1920 with a sound chemical base, and these were maintained with little change until a biochemical department was added at the end of 1941 to handle plant and animal nutritional problems more effectively. Easterfield's profound belief in research was applied as vigorously with his colleagues at *The Cawthron* as it had been in Wellington and a voluminous quantity of research was published during its first decade leading to the international reputation that the Institute gained.

From the turn of the century, second-class back blocks of land had been brought into production in the area, not least in the immediate post-war years. One of the early needs came from the pipfruit growers to improve their soils and to rid their crops of woolly aphis, bitter pit, black spot, and codling moth [23, 24]. Tillyard went to the UK to learn about woolly aphis and persuaded Easterfield to acquire one of the best entomological libraries then available in England. Kathleen Curtis attacked the black spot problem, while Rigg attended to the poor Moutere soil. Their combined studies were of immense value to the industry and were seen in improved yields and quality of the fruits. The woolly aphis problem was essentially solved by 1930, when spraying of apple trees was no longer needed following introduction of the parasitoid wasp *Aphelinus mali*, the elucidation of the life history of the black spot fungus of apples led to its eradication, and the fertiliser requirements of the Moutere Hills (and other soils) were solved or, at worst, improved in other areas. Furthermore, the Cawthron identification of the major causes of apple storage defects were of great value not only to the apple industry of Nelson but to fruit culture throughout New Zealand. The impact of the Cawthron work had the whole of the fruit growing business back in boom by 1926 when the crop output exceeded the shipping capacity. Apple exports soared from 6000 to 227,000 cases, and by 1930 the acreage was so high and the demand for seasonal pickers so great that there was no accommodation left available to them. The tobacco industry also prospered with the number of growers in Waimea County rising from 150 to 700 between 1926 and 1933.

As director of the Cawthron Institute, Thomas Easterfield played an important role in its establishment and in the development of research programmes. The early phenomenal success of the Institute was due in large measure to his enthusiasm, broad vision, and sound judgment in selecting the pioneer staff. But, as for all research establishments, funding was an on-going concern. The requirement of stamp duty on the Cawthron bequest reduced the interest returns and the continuing existence of the Institute depended on grants, donations and bequests. The lack of funding expressed itself as early as 1923 when Thomas Easterfield paid the salary of one his staff from his own pocket because there were inadequate monies available. The lack of funding was felt most strongly by Easterfield from his inability to undertake pure research in his own time. Monies were subsequently boosted following the 1925 visit to Nelson of Sir Ernest Rutherford and then the 1926 visit of Sir Francis Heath that rapidly led to the establishment of the New Zealand DSIR.

Easterfield's researches at the Cawthron led to ten publications, three on mineral oils of New Zealand and two on the occurrence of xanthin (**4**) [a close relative of caffeine (**5**); Fig. 9.2) in sheep, all of a technical nature. What Easterfield did with aplomb was to gain the trust of the industries and to provide them with relevant pamphlets and other appropriate educational aids, and become a friend of the local farmers, growers and industrialists, as well as a noted figure in the Nelson community. To his staff he was a man to be emulated. Elsa Kidson, a staff member at the Cawthron and first female demonstrator in a New Zealand University, recounted in an address given to the P. W. Robertson Society (for Victoria's chemistry students and graduates established in 1970) on April 9, 1974 [2, 25]: *I found him able to instil confidence and make me feel more capable, I am sure, than my abilities justified ..... but I can say that I never heard a word of criticism against him. He was always courteous and I never saw him get angry.* He was easily approachable, had a great sense of humour, and was a great reader in his spare time.

In the first Cawthron Lecture, Easterfield had said:

> I foretell a brilliant future for the Institute. The problems solved in it will lead to results of the greatest value to this city, the Dominion and to the human race. And the Institute itself is destined to become a centre of light, learning and culture honoured throughout the civilised world and a lasting tribute to Thomas Cawthron.

The results of his labours as Director of the Institute show that not only was Easterfield a dedicated founding chemist in this country, but that his ability to tell the future also was accurate.

In 1933 on the eve of his retirement, he delivered the annual Cawthron lecture that was entitled *The Thomas Cawthron Centenary Lecture* in which he summarised the successes of the Institute from its inception to that date [26]. Because of the role the Cawthron Institute played in solving the district fruit and dairy problems, the Nelson area was not impacted upon to the same extent as the rest of this country during the Great Depression of the 1930s. Moreover, the establishment of the McKees Fruitgrowers Chemical Company from 1931 in an abandoned coolstore in Mapua next to the wharf made the needed chemicals easily available in that era when chemical aids were regarded more favourably than now [24]. The successes of the Cawthron Institute over the early years are described by Miller in his 1963 book *Thomas Cawthron and the Cawthron Institute* [27] and the reader may wish to refer to it.

Thomas and Anna Easterfield had five children, two daughters before they arrived from England and then two more daughters (Muriel Helen; February 26, 1900 and Theodora Clemens; June 12, 1902) and a son (Thomas Edward; 1912) all of whom distinguished themselves. The two New Zealand born daughters became medics whilst Edward gained an M.Sc. from Canterbury University College in 1934 (following private school education in Wellington and Nelson) and subsequently an overseas Ph.D. and a career becoming a Principal Scientific Officer in the DSIR in London. Thomas Hill Easterfield died in Nelson on March 1, 1949, survived by his wife and children. He made an outstanding contribution to science in New Zealand. Remembered for his high spirits and cheerfulness, he set high standards in the training of students and in the conduct of chemical research. He established two notable institutions in Victoria's chemistry department and the Cawthron Institute. It was his vision, judgement and enthusiasm for research that set chemistry in Wellington on track and brought the Cawthron to the international position it was to hold at his retirement and maintains today.

Thomas Easterfield was appointed to the New Zealand Order of Merit in 1925, was awarded a King George V Silver Jubilee Medal in 1935, and knighted (KBE) in 1938. He was a Fellow of the New Zealand Institute that became The Royal Society of New Zealand and its 1921–1922 President, a Fellow of the NZIC, and a member of many other chemical societies.

References

1. Davis BR (1996) Easterfield, Thomas Hill. Dictionary of New Zealand biography. Te Ara—the encyclopedia of New Zealand. https://teara.govt.nz/en/biographies/3e1/easterfield-thomas-hill. Accessed 25 Jan 2019
2. MacFarlane DR (1978) T. H. Easterfield: Science in Colonial New Zealand. B.Sc. Honours Report, 18 leaves. Victoria University of Wellington Library Q143 E13 M143
3. Mackay D (2011) An appetite for wonder—Cawthron Institute 1921–2011, Cawthron Institute, pp 35–56

4. Easterfield TH (1883) The interaction of solutions of alum and sodium thiosulphate. Leeds Photographic Society
5. Easterfield TH (1883) A Glacial Deposit near Doncaster. Yorkshire Geological Society 8:212–213
6. Rigg T (1966) Easterfield, Sir Thomas Hill, K. B. E. In: McLintock AH (ed) An encyclopaedia of New Zealand, Te Ara—the encyclopedia of New Zealand. http://www.TeAra.govt.nz/en/1966/easterfield-sir-thomas-hill-kbe. Accessed 26 Sept 2016
7. Marsden E (1952) Obituary notices: Thomas Hill Easterfield, 1866–1949. J Chem Soc 1557
8. Miller D (1949) Obituary: Sir Thomas Hill Easterfield, K. B. E. Nature 163:669
9. Askew HO (1950) Obituary: Thomas Hill Easterfield (1866–1949). Trans Proc Royal Soc NZ 78:381–383
10. Easterfield TH, Sell WJ (1893) Studies on citrazinic acid. Part 1. J Chem Soc Trans 63:1035–1051
11. Easterfield TH, Sell WJ (1894) Studies on citrazinic acid. Part 2. J Chem Soc Trans 65: 28–31
12. Easterfield TH, Sell WJ (1894) Studies on citrazinic acid. Part 3. J Chem Soc Trans 65:828–834
13. Easterfield TH (1949) Spike, The VUC Review Golden Jubilee Number. 19–22 May
14. Beaglehole JC (1949) Victoria University College—an essay towards a history. NZ University Press, Wellington
15. Halton B (2018) Chemistry at Victoria—The Wellington University, 3rd ed. Victoria University of Wellington, Wellington. Available for free download at https://www.vic-toria.ac.nz/scps/about/attachments/chemistry-at-victoria-third-edition.pdf
16. Barrowman R (1999) Victoria University of Wellington, 1899–1999: a history. Victoria University Press, Wellington
17. Easterfield TH (1899) In: Inaugural addresses of the Victoria University College. Turnbull, Hickson & Palmer, Wellington, pp 36–43
18. Easterfield TH (1924) The development of science at Victoria College. Spike, The VUC Review, Easter, pp 44–47
19. Easterfield TH, Aston BC (1900) The Tutu Plant Part 1. Trans N Z Inst 33:345–355
20. Easterfield TH, Aston BC (1900) Tutu. Part I. Proc Chem Soc 16:211–212
21. Easterfield TH, Aston BC (1901) Tutu. Part I. Tutin and coriamyrtin. J Chem Soc Trans 79:120–126
22. Soper FG (1975) Chemistry in New Zealand. Chem N Z 39:97–101
23. MacKay D (2011) An appetite for wonder: Cawthron Institute 1921–2011. Midas Printing (China) for the Cawthron Institute
24. McAloon J (1997) Nelson: a regional history. Cape Catley in association with Nelson City Council, Whatamango Bay, New Zealand, pp 157–163
25. Burns GR, Duncan JF, Shorland FB (1974) T. H. Easterfield on the 75th anniversary of the Founding of the Chemistry Department, Chemistry Department Report No. 4, 9 April 1974, VUW library QD1 V645 R 4
26. Easterfield TH (1934) The achievements of the Cawthron Institute. The Cawthron Institute, Evening Mail Office, Nelson, Victoria University of Wellington library AS750 NEL C
27. Miller D (1963) Thomas Cawthron and the Cawthron Institute, Cawthron Institute, pp 95–136 give a summary to the important work of the Institute over the earlier years

Chapter 10
Sir Edward Frankland (1825–1899)

The name of Edward Frankland (Fig. 10.1) is recognised by many chemists, but his contribution to the discipline and profession remembered by few. He was commemorated by the Royal Society of Chemistry in 1984, initially with the Sir Edward Frankland Fellowship, now the Frankland Award. It recognises study in organometallic chemistry—but what was it that Frankland did and why is he so recognised?

Edward Frankland was born near the small town of Garstang in Lancashire, England. The town lies between Preston and Lancaster and is east of Blackpool [1, 2]. Edward was the son of Margaret (Peggy) Frankland who went into service with the wealthy Gorst family in Preston in 1824. She had an affair with the young heir, Edward Chaddock Gorst and, when discovered, the horrified family dispatched the pregnant Peggy with a handsome annuity, providing that the identity of the father was never disclosed. It was revealed only by 20th century historians. Peggy moved to Churchtown, in the Parish of Garstang, where her son Edward was born on 18th of January 1825. He was accorded her family name but given the first name of his father.

Edward Frankland's childhood was ferocious as he and his mother were outcasts. Nonetheless, she ensured that Edward got an education using the annuity to meet the costs [1]. By age seven he had been to seven schools where, it appears, he was routinely beaten. It was only in his eighth school that he gained an element of stability as he stayed there until he was twelve [3]. This school, in Lancaster, was an enlightened one, as the pupils were encouraged to acquire first-hand knowledge of nature and even conduct simple experiments. From twelve to fifteen years of age the young Edward attended the Lancaster Free Grammar School, although this time is recorded in the Oxford Dictionary of National Biography as 'entirely unproductive' [4]. By then, Peggy had married William Helm, one of the first lodgers at the Lancaster guesthouse that she had run. At 15, Edward's wish was to enter the medical profession

---

A previous version of this chapter was published as Halton B (2013) Sir Edward Frankland KCB, FRS, FCS (1825–1899). Chem N Z 77:92–96.

Fig. 10.1 Left: Edward Frankland; right: Adolph W. H. Kolbe (A. Brasch, Leipzig) (both out of copyright from Wikipedia)

and so, on advice from others, his stepfather William Helm apprenticed him to a local druggist (pharmacist) by the name of Stephen Ross.

It seems that the apprenticeship was one of drudgery: Ross appears to have taught him little, requiring 70-hour weeks of bone-breaking labour, hauling, wrapping and grinding [1]. In his writings, Frankland speaks of his apprenticeship as wasted time, calling it, "six years' continuous hard labour, from which I derived no advantage whatever, except the facility of tying up parcels neatly" [5]. Nonetheless, it is probable that he did gain some useful skills, not least the safe handling of chemicals. Despite the long working hours, Edward used his limited free time to borrow books from the mechanics' institute, attend the classes and conduct some experiments in the makeshift laboratory that the local doctors, Christopher Johnson and his son James, had made available there [4]. Apparently, they evicted a tenant from a cottage to create the institute for local lads. Subsequently, Dr. James Johnson was instrumental in gaining Frankland a position at the end of his apprenticeship late in 1845 with Lyon Playfair, the then recently appointed chemist in the Government Department of Woods and Forests in London. Frankland's progress there was so fast that, after six months, became as Playfair's lecture assistant for the additional post of Professor at the Putney College of Engineering that Playfair had taken. Apart from being assistant, Frankland took the course that Playfair gave and successfully completed the final examination, the only one he ever sat [4].

One of Playfair's other assistants at that time was Herman Kolbe (Fig. 10.1), who had gained a Ph.D. with Bunsen in Marburg and come to the belief that organic compounds were comprised of identifiable groups of atoms which he termed radicals (then spelled 'radicles'). Frankland became caught up in this and, with Kolbe, showed that the hydrolysis of cyanoethane led to propionic acid, thus confirming the hypothesis. The results were communicated to the Chemical Society in April, 1847. The following month saw the pair go to Marburg where Frankland spent some time in

10 Sir Edward Frankland (1825–1899)

(a)

[Structure of kyanethine: pyrimidine ring with H₂N at position 4, Et at positions 2 and 6, Me at position 5] ←—K— CH₃CH₂CN ——→ CH₃CH₂CO₂H

1, kyanethine : 2,6-diethyl-4-methylpyrimidinamine

(b) 2EtI + Zn ——→ Et₂Zn + I₂ ; 2MeI + Zn ——→ I₂ + Me₂Zn $\xrightarrow{2H_2O}$ 2MeH + Zn(OH)₂

Fig. 10.2 **a** Transformations of cyanoethane and **b** synthesis of diethyl zinc

Bunsen's laboratory attempting to prepare the ethyl radical by dropping cyanoethane onto potassium metal. In his words, "a very vigorous reaction often accompanied by fire and rapid gas evolution" took place [6]. The gaseous product was not the desired ethane but butane, which required the cyanide to be impure and contain some water. A side product was also obtained, isolated, purified, analyzed and named kyanethine. It was subsequently proved to be the pyrimidine **1**, a trimer of cyanoethane (Fig. 10.2a) [6].

Frankland stayed for only three months in the Bunsen laboratory, as he was appointed a teacher at Queenswood School in King's Somborne, west of Winchester in Hampshire. He took up his post in September of 1847 and found that one of the other teachers was John Tyndall (Fig. 10.3), who became a prominent 19th century English physicist. The pair taught one of the first laboratory-based science courses in England, each teaching the other his own subject by rising at 4 am to exchange lessons before schoolwork began [7]. Again Frankland was drawn to research, and he continued his quest for the ethyl radical as he knew it. Initially, he tried to remove the oxygen atom from diethyl ether using both potassamide (potassium amide and potassium metal, but was unsuccessful. He then directed his attention, not to cyanoethane again, but to the iodide since HI was known to easily be decomposed by potassium. After an explosion with potassium metal, he switched to zinc, heating the reagents in a sealed tube on 28 July 1848. Unfortunately he had no means of measuring or estimating the gaseous product formed in the sealed tube since his eudiometer he had had been broken (a eudiometer is similar to a graduated cylinder but closed at the top end with the bottom immersed in water or mercury thereby allowing a gas to be collected in the cylinder and its volume measured; see Fig. 3.10). Frankland left the tubes unopened, resigned his post leaving after a year to return to the Marburg laboratory, this time with Tyndall and the sealed tubes.

His serious research started immediately during the summer of 1848. He studied the action of sodium on fatty acids, obtaining radicals such as 'methyl', which, in reality, was ethane. Opening one of the sealed tubes that he had brought from Hampshire under water in mid-February 1849 led to an evolution of gas that was apparently as spectacular as it was significant—he obtained ethyl (in reality butane)! His researches were successfully submitted for the Ph.D. degree in July, 1849. During his last weeks with Bunsen he noticed a liquid product in one of his unopened

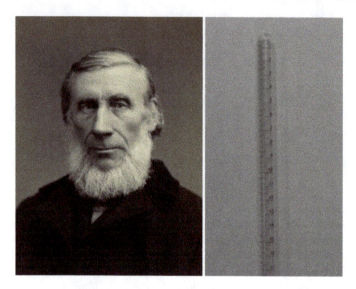

Fig. 10.3 Left: John Tyndall mid-career (out of copyright.wikimedia). Right: the closed end of a eudiometer (Skiaholic at English Wikipedia)

Queenswood tubes from the reaction of ethyl iodide with zinc. This liquid contained zinc, but it was an organic compound and it contained the metal! It was diethyl zinc as we now know it (Fig. 10.2b). Organometallic chemistry was born with this July 1848 experiment.

On 12 July 1849 in Bunsen's laboratory he prepared dimethyl zinc (Fig. 10.2b) by digesting methyl iodide with zinc. After discharging the gasses, he cut off the upper part of the tube to test the action of water on the solid residue. On adding water he observed [8], "a greenish-blue flame several feet long shot out of the tube, causing great excitement amongst those present and diffusing an abominable odour through the laboratory".

Frankland moved from Marburg to Liebig's laboratory in Giessen for a few months then moved back to England early in 1850 as Professor of Practical Chemistry at Putney College [9]. He remained there for just one year, but it was a year full of research and significant output. During this time, he examined the action of light on reactions between metals and alkyl halides, taking advantage of a flat part of the roof of his chemistry laboratory and the increasing daylight of the spring. The outcome was the formation of a range of organometallic compounds involving zinc, tin, mercury, etc., that firmly established the new branch of chemistry. At least as important were the deductions he made from his new data. He recorded regularity in the formulae of his compounds and came to recognise that each element had a limit to the number of radicals that could be attached to it; that each had a definite combining power. Thus, from his novel yet obscure compounds, Frankland discovered one of the most significant principles of chemistry, namely that known as valency, and he was the first to articulate the concept by terming it "combining power". The idea of valency laid the groundwork for Kekule's hypothesis of the structure of benzene and for Gerhardt's theory of types. The results, communicated to the Royal Society as his second paper on organometallic compounds and read at a meeting on 10 May 1852, were published subsequently [10, 11], securing his place in the documented history of the valency concept [12].

On 2 January 1851, Frankland was appointed as the inaugural professor of chemistry at Owens College in Manchester, but before he left London he married Sophie Fick whom he had first met while with Bunsen in Marburg in 1847. The objective of the new Owens College was the provision of instruction and improvement to young men in such branches of learning and science as are usually taught in the English Universities; it was the beginning of Victoria University of Manchester into which it evolved [13]. During his six and a half years at the college, Frankland firstly designed the laboratories, established his own research using a new Mancunian technique of heating under pressure in an iron digester, and became consultant to a range of industries to a scale unprecedented among other universities. This led to Manchester's tradition in applied chemistry. His researches led to several new products and allowed for the synthesis of the zinc alkyls in large quantity. He was elected to the Fellowship of the Royal Society in 1853 and was awarded its Royal Medal in 1857.

Despite Frankland's reputation, Owens College was mismanaged and its students were of indifferent quality; it almost closed [4, 13]. So it was, that in mid-1857, he accepted the position of lecturer in chemistry as St. Bartholomew's Hospital and moved to London with his wife and, by then, three children. While he had little opportunity for research there, he was able to use the better facilities of the Royal College of Chemistry in Oxford Street, headed by A. W. Hofmann. That college became the Royal School of Science in 1900, was the first part of Imperial College, and eventually became its Chemistry Department. Despite the Royal College facilities, Frankland took other positions that continued to make his life hectic, yet his research output surged.

Michael Faraday retired from the Royal Institution in 1861 and Frankland was appointed, initially on a temporary basis. Once tenured, he stayed there until 1868. However, from 1865 he acted as replacement for Hofmann at the Royal College while the latter was on three years' leave in Germany. Although this position became permanent in 1868, when Hofmann elected to remain in Germany, Frankland resigned from this post within the year.

The early part of Frankland's time in London saw him expand his interests by carrying out research in such diverse subjects as illuminating gases, explosives, atmospheric chemistry, and water quality. He also became involved with the Royal Society and the Chemical Society, serving on their Councils. He was a member of the influential X Club from its inception by Thomas Henry Huxley in 1864 until its closure in 1893. The X Club was a dining club of nine men who supported the theories of natural selection and academic liberalism in late 19th-century England.

During his time as replacement for Hofmann, he was required to assess the examination papers set by the Department of Science and Art, which led to him to publishing his *Lecture Notes for Chemical Students* [14]. Here, the atoms were represented by their letters and joined with 'bonds' (a term introduced in the book) and the concepts of valence were discussed. His involvement with educational matters led him to espouse the need for laboratory training in chemistry for all students of the subject, and from 1869, he provided a lab course for interested teachers free of charge.

Back in London, Frankland's initial studies extended his organometallic work and, from 1859 with Duppa, he carried out the first studies on organoboron compounds that were published from 1860 and included the very reactive trialkyl-borons [15–17]. Not only was work on organomercury and phosphorus compounds reported, but in addition, Frankland and Duppa made significant contributions to synthesis of ethers, esters and dicarboxylic, hydroxy and unsaturated acids, revealing the structure and relationship between the compounds, all coming from the exploitation of their organometallics. To some, Edward Frankland is regarded as a founder of synthetic organic chemistry.

Despite Frankland's reputation, Owens College was mismanaged and its students were of indifferent quality; it almost closed [4, 13]. So it was, that in mid-1857, he accepted the position of lecturer in chemistry as St. Bartholomew's Hospital and moved to London with his wife and, by then, three children. While he had little opportunity for research there, he was able to use the better facilities of the Royal College of Chemistry in Oxford Street, headed by A. W. Hofmann. That college became the Royal School of Science in 1900, was the first part of Imperial College, and eventually became its Chemistry Department. Despite the Royal College facilities, Frankland took other positions that continued to make his life hectic, yet his research output surged.

Michael Faraday retired from the Royal Institution in 1861 and Frankland was appointed, initially on a temporary basis. Once tenured, he stayed there until 1868. However, from 1865 he acted as replacement for Hofmann at the Royal College while the latter was on three years' leave in Germany. Although this position became permanent in 1868, when Hofmann elected to remain in Germany, Frankland resigned from this post within the year.

The early part of Frankland's time in London saw him expand his interests by carrying out research in such diverse subjects as illuminating gases, explosives, atmospheric chemistry, and water quality. He also became involved with the Royal Society and the Chemical Society, serving on their Councils. He was a member of the influential X Club from its inception by Thomas Henry Huxley in 1864 until its closure in 1893. The X Club was a dining club of nine men who supported the theories of natural selection and academic liberalism in late 19th-century England.

Despite all of the foregoing, Frankland's international reputation came just as much from work in applied chemistry. It was noted above that he had worked on water quality. In fact, it was during the summer of 1859 that he assisted Hofmann in reporting to the Metropolitan Board of Works in London on possible means of deodourizing sewage. At that time it was sent raw into the River Thames making it black and horribly offensive, causing much water-borne disease [7]. This involvement with water analysis and water purification gained momentum after he took over Hofmann's role at the Royal College. He found that he had inherited the role as analyst for the London water supply, and was appointed to the second royal commission on the pollution of rivers in 1868. This also provided him with a fully equipped government laboratory, which led to much valuable scientific information on sewage contamination, industrial effluent and water purification for domestic consumption over a six-year period. Within a very short time (with the assistance of H. E. Armstrong) he had developed a new analytical method for carbon and nitrogen in water. Later he suggested that previous sewage contamination could be detected from nitrate concentration. He pursued relentlessly the quest for safe drinking water, campaigning against supplies that failed to meet his standards. While this led to antagonism from many chemists employed by water companies, he became recognized as the world leader in the field. Increasingly, he was asked to analyze water samples to the extent that his consultancy demanded more and more of his time. He was analyzing water samples from numerous international clients. This work moved to a privately funded laboratory from about 1862 and continued well into his retirement. It has been said that he did not take kindly to the criticism that his analytical work took him away from his role at the Royal College, and that it was this that persuaded him into early retirement in 1885 aged 60 years.

The increasing recognition that Frankland attracted in the 1870s led to his reappointment to the Council of the Royal Society and the office of Vice-President, and then the 1871–73 President of the Chemical Society. His great concern for the education of chemists, the provision of laboratory programmes and the needs of the profession led him to campaign for professionalism in the discipline. So successful was he that in 1877 the Institute of Chemistry of Great Britain was founded, with a focus on the qualifications and professional status of chemists. Its aim was to ensure that consulting and analytical chemists were properly trained and qualified. It was the first of the professional organizations for scientists to be created, and Frankland served as its first President from 1877 to 1880. New Zealand Institute of Chemistry, this author's own country's Institution owes its origins, with the professional statutes and status that it accepted from its inauguration in 1931 until its move to become a learned society from the early 1990s.

The 1901 supplement to the *Dictionary National Biography* [7] states that Frankland published sixty-three papers alone, fifteen with B. F. Duppa, three with J. N. Lockyer (leading to Lockyer's recognition of helium in the sun's atmosphere), two with H. Kolbe (though a further one omitted his name), one with H. E. Armstrong, and ten with other chemists. Forty-nine of these are recognised by SciFinder® though several of the earlier papers carry no year. The Royal Society archival journals carry forty papers by Frankland that date from 1850. However, the *Complete Dictionary of Scientific Biography* [18] states that Frankland published over 130 papers of which the *Royal Society Catalogue of Scientific Papers* (London 1867–1925) lists 107. Irrespective of the actual number, Edward Frankland made a major contribution to the subject of chemistry, to the nature of the profession, and to the well-being of mankind. In retirement he received the Royal Society's Copley Medal (1894) and a KCB in the Queen Victoria's Diamond Jubilee Honours of 1897 for his water quality work.

References

1. Leinhard JH (2005) No. 2036: Edward Frankland. Engines of Our Ingenuity. www.uh.edu/engines/epi2036.htm. Accessed 25 Feb 2013
2. Graham GH (2016) Sir Edward Frankland, 1825–1899. http://ghgraham.org/edwardfrankland1825.html. Accessed 25 Jan 2019
3. McLeod H (1905) Obituary, Edward Frankland. J Chem Soc Trans 87:565–618
4. Russell CA Frankland, Sir Edward. Oxford Dictionary of National Biography. www.oxforddnb.com/view/article/10083?docPos=1. Accessed 28 Feb 2013
5. Eric (2010) Edward Frankland. Dead scientist of the week. http://deadscientistofthe-week.blogspot.co.nz/2010/01/edward-frankland.html. Accessed 28 Feb 2013
6. Dingers J, Lamberth C (eds) (2012) Bioactive heterocyclic compound classes: pharmaceuticals. Wiley-VCH, pp 287–288
7. Harthog PJ (1901) Frankland, Edward (DNB01). Dictionary of National Biography, 1901 Supplement. http://en.wikisource.org/wiki/Frankland,_Edward_(DNB01). Accessed 7 Mar 2013
8. Frankland E (1877) Experimental researches in pure, applied, and physical chemistry. John Van Voorst, London
9. Walford E (1878) Putney. In: Old and New London: Volume 6. Cassell, Petter & Galpin, London, pp 489–503. British History Online. http://www.british-history.ac.uk/old-new-london/vol6/pp489-503. Accessed 4 Mar 2013
10. Frankland E (1852) On a series of organic bodies containing metals. Phil Trans R Soc Lond 142:417–444
11. Frankland E (1861) On organo-metallic bodies: a discourse delivered to the members of the Chemical Society of London. Quart J Chem Soc 13:177–235
12. Russell CA (1971) The history of valency. Leicester University Press, Leicester
13. Hartog P (1900) The Owens College, Manchester (founded 1851); a brief history of the college and description of its various departments. Cornish, Manchester
14. Frankland E (1866) Lecture notes for chemical students: embracing mineral and organic chemistry. John Van Voorst, Parternoster Row, London
15. Frankland E, Duppa B (1859) On boric ethide. Proc R Soc Lond 10:568–570. (Volume 10 of the Proceedings of the Royal Society is dated 1859, but this paper was submitted in 1860)
16. Frankland E (1860) On a new series of compounds containing boron. Proc R Soc Lond 12:123–128
17. Frankland E (1862) On a new series of organic compounds containing boron. Phil Trans R Soc Lond 152:167–183
18. Brock WH (2008) Frankland, Edward. Complete Dictionary of Scientific Biography. www.encyclopedia.com/topic/Edward_Frankland.aspx. Accessed 11 Mar 2013

Chapter 11
William Henry (1774–1836)

William Henry (Fig. 11.1) was born on December 12 in 1774 at 19 St. Ann's Square, Manchester. He was the third son of Thomas Henry and his wife Mary (née, Kinsey) [1–3]. The Henry family hold the distinction of having three generations of chemists, Thomas (1773) [1], William (1809) [4] and William's son (William) Charles (1834) [5], holding Fellowships of the Royal Society continuously for close on one hundred and twenty years from 1773 [5].

Henry senior (1734–1816) was a surgeon and apothecary (a trading chemist) who taught himself chemistry from the traditional eighteenth century text *Elementa Chemiae* written in 1732 by Dutch chemist Boerhaave. Although a surgeon, Thomas Henry was not an MD as, in those days, the majority of surgeons qualified by apprenticeship through an apothecary. However, it was Thomas Henry's chemical acumen that placed him among the more affluent members of Lancashire society. This came from the manufacture of magnesia—magnesium oxide—for which he devised an industrial process and gained the name *Magnesia Henry*. Although little is known about the education of his elder sons, Thomas and Peter, William was privately educated by the Rev. Ralph Harrison who taught Latin and Greek at the nearby Unitarian Cross Creek Chapel, the Dissenters' Meeting House. When the Manchester Academy (which Thomas Henry had helped establish) opened in 1786, Harrison was appointed Professor of Classical Literature [3, 4] and William, although only eleven years old and below entry age, was permitted to follow his tutor there. The academy was run by English Presbyterians as one of several dissenting academies that provided religious nonconformists with higher education—the English universities of Oxford and Cambridge took only Anglicans. As a young boy, William suffered serious injury when a heavy beam fell on his right side. The consequential acute neurological pain he suffered remained with him throughout his life and turned him to study since normal physical boyhood activity was limited. In 1790, at the age of sixteen, he left the academy to become secretary-companion to Thomas Percival, a colleague of his

---

A previous version of this chapter was published as Halton B (2014) William Henry, MD, FRS (1774–1836). Chem N Z 78:128–131.

© Springer Nature Switzerland AG 2020
B. Halton, *Some Forgotten Chemists*, Perspectives on the History of Chemistry,
https://doi.org/10.1007/978-3-030-16403-4_11

Fig. 11.1 Left: William Henry by James Lonsdale, detail of an engraving by Henry Cousins after a portrait by James Lonsdale (courtesy of the trustees of the British Museum from Wikipedia); right: John Dalton adapted by John Price Millington (author), James Stephenson, artist, 1828—1886 (Wikimedia Commons)

father and the leading physician in Manchester, perhaps best known for crafting the first modern code of medical ethics published in 1794. Percival had poor eyesight and suffered violent headaches, and William's job was to read aloud to him, keeping him familiar with developments in medicine and science, and then taking whatever dictation was required. Thus, William Henry became familiar with Percival's correspondence with the noted men of science and literature of the day. During his time with him, William began to study medicine and he entered Edinburgh University for a medical degree in the winter of 1795.

Edinburgh was the centre of modern medical education in the UK as Oxford and Cambridge held strong to classical medical tradition. While there, the young Henry attended lectures in chemistry given by Joseph Black (1728–1799) who, though old and frail, was still the Professor of Chemistry [4]. He became more drawn to science than medicine and he performed his first piece of serious scientific research there—studies of carbonated hydrogen gas. This was read to the Royal Society by his father on June 29, 1797 [6]. After a year of medical school, Henry senior recalled William to Manchester to help run the family businesses as his elder brothers had little appetite for business and lacked the aptitude to assist. William became central to the family affairs [1, 5], and after a short time, he was taken into partnership by his father. He ran the magnesia factory, the mainstay of the family fortune, under the name T & W Henry from 1797. It survived under this name until the end of 1933. Late in 1805, William returned to Edinburgh to complete his studies, leaving the factory in the hands of a manager. His two years there gave him his only period freed from commercial responsibilities and he graduated MD with a thesis on uric acid (*De acido urico et morbis a nimia ejus secretion ortis*), a substance that continued to hold his interest.

$$H_2 + Cl_2 \longleftarrow 2H^\bullet + 2Cl^\bullet \longleftarrow 2HCl + 2Hg \longrightarrow Hg_2Cl_2 + H_2 \xrightarrow{\frac{1}{2}O_2} H_2O$$

Fig. 11.2 Henry's electrical discharges of HCl

William Henry had a natural talent for experimental study and it was over the ten years following his first return from Edinburgh that he made his major contributions to chemistry, carrying the skills he learned as a manufacturing chemist to the research bench. Apart from magnesia, the other major activity of the Henrys was the production of aerated waters—soda water with or without added flavouring. Thus, William Henry had a life-long interest in gases, their essential properties and their chemical behaviour, and, later, he became actively involved in the gas lighting industry in Great Britain. His interest in gases led him to work with his father on pneumatic medicine—the inhalation of gases to treat disease and especially consumption—at the Royal Manchester Infirmary and it directed much of his early work. In this, William studied the composition and decomposition of muriatic acid gas (HCl) which, like all acids at that time, was thought to contain oxygen. He came close to solving the problem of its composition in 1800, some ten years ahead of Humphrey Davy. His results appeared in the *Philosophical Transactions of the Royal Society* in 1800 [7]. Henry repeatedly exposed HCl to electric discharges and when performed over mercury he saw a volume reduction and the formation of a white solid [now recognised as mercury(I) chloride] (Fig. 11.2).

When repeated with HCl in the presence of O_2, a greater volume drop was seen as water is formed from reaction of the liberated hydrogen with the oxygen, and in the absence of mercury, chlorine was produced (Fig. 11.2). When Davy finally showed that muriatic acid was comprised of hydrogen and chlorine only, Henry supported him and provided additional evidence in 1812 [8]. Although Henry was unable to come to the correct conclusion regarding HCl until after Davy's paper, it is clear that his results are correct and of significance. However, the work for which he is best known is on the solubility of gases.

Henry's studies on gas solubility gave rise to what we now regard as Henry's law. It also led to a friendship and collaboration with the teacher John Dalton, most notably from 1800 to 1805 [9]. He helped the colour-blind Dalton with his experiments and, while Dalton subsequently became world-famous for his theories, his practical abilities were less enduring. Thus, Dalton and Henry shared an interest in the chemistry of gases and liquids. Understandably, William Henry approached his study of them from the viewpoint of an industrialist who manufactured soda water and hoped to use gases in medicine. Dalton, on the other hand, came to chemistry from a metrological background.

By about 1800 Dalton saw the atmosphere as comprising four types of particle— the atoms of oxygen and nitrogen and the compound atoms of water and carbonic acid (carbon dioxide) that were motionless. He could not understand why a puddle of water could diffuse into the atmosphere or why the air did not separate into layers. His discussions with Henry as to what might cause these effects were probably important in the development of his atomic theory. In 1801 Dalton formulated the concept that there was a repulsive force between particles of the same kind in a gas [10, 11], viz. like repels like, and that this resulted in every particle of water in air getting as far from another as possible so that the whole would be evenly distributed throughout the available space. William Henry, initially opposed to this, converted to acceptance by 1804 saying every gas is a vacuum to every other gas [12, 13]. The change stemmed from his and Dalton's experiments on gas solubility, notable over 1801 and 1802 when both worked on the solubility of gases in water. Because most previous study had been with carbonic acid (carbon dioxide), Henry chose this as his first target and he reported his results to the Royal Society (London) just before Christmas in 1802 and published them in the *Philosophical Transactions* early in 1803 [14]. Dalton read his studies to the Manchester Literary and Philosophical Society some ten months later (October 23, 1803), and published them in the Memoirs of the Literary and Philosophical Society of Manchester in 1805 [15] Henry's conclusion, which appears in an appendix to the original paper [14], became known as Henry's law and for this he was awarded the Copley Medal in 1808. The law, one of the fundamental gas laws, states:

> At a constant temperature, the amount of a given gas that dissolves in a given type and volume of liquid is directly proportional to the partial pressure of that gas in equilibrium with that liquid.

An equivalent is:

> The solubility of a gas in a liquid is directly proportional to the partial pressure of the gas above the liquid.

Not surprisingly, the most common practical illustration of Henry's law is provided by carbonated soft drinks. Before the container of carbonated drink is opened, the gas above the drink is almost pure carbon dioxide at a pressure slightly higher than atmospheric. The drink itself contains dissolved carbon dioxide. When the bottle or can is opened, some of this gas escapes, giving the characteristic hiss (pop in the case of sparkling wines). Because the partial pressure of carbon dioxide above the liquid is now lower, some of the dissolved carbon dioxide comes out of solution as bubbles. Obviously, when a glass of the drink is left in the open, the concentration of carbon dioxide in solution equilibrates with the carbon dioxide in the air, and the drink goes flat. A more complex example of Henry's law is in the bends that can be suffered by underwater divers.

Dalton's experiments on the solubility of gases provided a mechanical model of it and led to his conclusion that the differences in the solubility of different gasses depended on the weight and number of the ultimate particles of the several gases [14], a generalisation that is now known as Dalton's law of partial pressures [15]. Henry was much more concerned with the facts associated with his studies than with theory, but when they fitted to a theory he was delighted. Thus, Henry began his study from the knowledge that the solubility of CO_2 in water was increased with increased pressure. However, in 1801, the preparation and handling of pure gases was essentially unknown and his work used gas mixtures—almost always containing amounts of air. This led to the overlap of his and Dalton's studies and Henry's 1802 presentation on the quantity of gases absorbed by water was possible only after taking into consideration the mixed gases and Dalton's law of partial pressures known to him but unpublished at that time. The nature of the experiments and the rather crude equipment then available has been nicely described by the Farrars and Scott [9] and does not justify further discussion here. Suffice to say that the true solubility of CO_2 could only be determined from an analysis of the undissolved gas and applying the law of partial pressures to it. This only became apparent to Henry after he had read his paper and presented his manuscript for publication. It was clarified in the appendix where he says that the absorption of gases by water is a purely mechanical effect and that the amount is exactly proportional to the density of the gas, independent of any other gas with which it may be mixed. The result of some 50 experiments on CO_2, H_2S, N_2O, O_2 and N_2 led to the formulation of Henry's law.

William Henry's subsequent work with gases included the electrolysis of ammonia in order to assist in determining its composition [1]. Here, a series of experiments led to the conclusion that dry ammonia doubled in volume (1 volume increased to 1.98 volumes) from electric discharge. We now know that two volumes of N_2 are replaced by four volumes of gas as per:

$$2NH_3 \rightarrow N_2 + 3H_2$$

However, it is his studies on the destructive distillation of coal and oil that led to his involvement with the gas lighting industry and it provided the only fundamental research in that industry until the mid-1800s [16]. His papers on this were published over the 1805–1821 period [17–21].

It was towards the end of the 18th century that the gaseous products from coal were beginning to receive attention. William Henry analysed the constituents of the gases produced and distinguished between some of them by chemical methods, and he studied their suitability for lighting. Thus, he showed that the gas mixtures comprised carbonic acid (CO), carburetted hydrogen (CH_4), hydrogen, olefiant gas ($CH_2 = CH_2$) together with some carbonic acid gas (CO_2) and sulfureted hydrogen (H_2S). He disagreed with other authors of the day in that he correctly showed the composition of the gases from coals, oils and other organic substances such as wood and peat. They comprised mixtures of a few simple compounds, predominantly hydrogen, methane and the oxides of carbon. Henry's last important paper on hydrocarbons [21] contained speculation on the way methane is formed "in natural operations". Although we know that water and charcoal yield hydrogen and the oxides of carbon (water gas), and that the formation of methane in stagnant pools is a microbial process, Henry's speculation provides one of the very early attempts to give a mechanism in terms of the atomic theory. One needs to remember that water was thought to be 'OH' and methane (carburetted hydrogen) 'CH_2' in reading that section of the original paper which appears below [21]. Not only does he account for the products as he saw them, he also proposes a metathetical way in which the 'OH' and 'C' approach and separate, a concept some 150 years ahead of olefin metathesis.

Beyond these insights, it was a result of William Henry's 1808 paper [18] that ethylene became easily identifiable. He showed that it forms an oily liquid (1,2-dichloroethane) on mixing with chlorine and it was from this and the 1795 work of Dutch chemists (Deimann, van Troostwyck, Lauwerenburgh and Bondt) that it became known as *olefiant gas*. This organic reaction was also well ahead of the understanding of halogen addition to multiple carbon-carbon bonds and some 20 years before Wohler's urea synthesis.

Henry's involvement with the gas lighting industry included experiments which showed that sulfureted hydrogen (H_2S) was the main contaminant of raw coal gas and that the best coals for illumination purposes unfortunately gave the greatest quantity of H_2S [16]. He suggested that the most effective way of removing the contaminant would be by agitation with quicklime and water. Thus, raw coal gas could be used for lighting only in well ventilated areas unless the hydrogen sulfide was removed. Samuel Clegg tried to put this into practise in a mill in Coventry in 1809 but it was unsuccessful because of the short time that the gas and lime were in contact. It seems that Clegg and Henry then collaborated in the installation of gas lighting at Stonyhurst College, the Jesuit College in Lancashire. This was the first non-industrial building to be so illuminated. For this they constructed a vessel containing lime-water through which the raw gas was bubbled prior to passing to the gas holder. Henry showed that the purified gas was perfectly free of the contaminant and the installation was a complete success. Lime was then used as a "sweetener" for the gas over the next fifty years [16].

William Henry's reputation was significant and he was invited to let his name be advanced for the inaugural Regius Chair in Chemistry at Glasgow University. He declined on the grounds of his family business, his health, and his growing family; he had nine children of whom six lived to maturity. The position went to Henry's friend from Edinburgh days, Thomas Thomson, who held it from 1818 until his death. William became a man of considerable wealth with homes to match. By the late 1820s his health was deteriorating and he took no further part in experimental chemistry, becoming an elder statesman whose opinions and views were sought and valued by many. He was elected a Fellow of the Royal Society in February 1809, having been awarded their prestigious Copley Medal in 1808. He held the position of Vice-Chairman of both the Literary and Philosophical Society and the Natural History Society of Manchester and was one of the founding members and life member of the British Association for the Advancement of Science. His 1799 textbook *An Epitome of Chemistry: In Three Parts* was renamed *Elements of Experimental Chemistry* and enjoyed considerable success, going through eleven editions over some 30 years [22].

After the marriage of his son Charles in 1834, William and the rest of his family moved to Pendlebury, some four miles out of Manchester. He died there some two years later on September 2, 1836. His neurological pains had increased to the extent that during the night he went to his private chapel and ended his life with a bullet through the mouth. His son, William Charles (who had qualified in medicine at Edinburgh in 1827, was an Honorary Physician at the Manchester Infirmary from 1828 and elected FRS in 1834), like his father before him, had been assisting with the magnesia factory and the family business increasingly as his father's health deteriorated. After his father's death he took over the company keeping the name and running it until his own death in 1892. However, after his father's death he retreated more and more from active participation, taking up the life of a country gentleman at Hatfield near Ledbury in Herefordshire. He also withdrew from practical science but kept many scientific friendships, notably one with Liebig who he had met in Germany and hosted on his first visit to Britain in 1837.

References

1. Thornber C. Thomas Henry, FRS and his son, William Henry, MD, FRS, GS. http://www.thornber.net/cheshire/ideasmen/henry.html. Accessed 22 Nov 2013
2. Henry WC (1837) A biographical account of the Late Dr. Henry. F. Looney, Manchester. Held in The Portico Library, 57 Mosley Street, Manchester, UK, M2 3HY (Ref. Fo18)
3. Farrar WV, Farrar KR, Scott EL (1973) The Henrys of Manchester Part I: Thomas Henry (1734–1816). Ambix 20:183–208
4. Farrar WV, Farrar KR, Scott EL (1977) The Henrys of Manchester. Part 6. William Charles Henry: The Magnesia Factory. Ambix 24:1–26
5. Farrar WV, Farrar KR, Scott EL (1974) The Henrys of Manchester Part 2. Thomas Henry's Sons: Thomas, Peter and William. Ambix 21:179–207
6. Henry W (1797) Experiments on carbonated hydrogenous gas; with a view to determine whether carbon be a simple or compound substance. Phil Trans Royal Soc Lond 87:401–415
7. Henry W (1800) Experiments for decomposing the muriatic acid. Phil Trans Royal Soc Lond 90:188–203
8. Henry W (1812) Aditional experiments on the muriatic and oxymuriatic acids. Phil Trans Royal Soc Lond 102:238–246
9. Farrar WV, Farrar KR, Scott EL (1974) The Henrys of Manchester Part 3. William Henry and John Dalton. Ambix 21:208–246
10. Dalton J (1801) New theory of the constitution of mixed aeriform fluids, and particularly of the atmosphere. Nicholson's J 5:241–244. (The Journal of Natural Philosophy, Chemistry, and the Arts was generally known as Nichol-son's Journal, an early scientific journal begun in 1797 in Great Britain by William Nicholson, the sole editor)
11. Dalton J (1802) Experimental essays, on the constitution of mixed gases; on the force of steam or vapour from water and other liquids in different temperatures, both in a torricellian vacuum and in air; on evaporation; and on the expansion of gases by Heat. Man Mem 5:535–606. (Memoirs of the Literary and Philosophical Society of Manchester is abbreviated to Man. Mem)
12. Henry W (1804) Illustrations of Mr. Dalton's Theory of the constitution of mixed gases. Nicholson's J 8:297–301
13. Henry W (1804) Illustration of Mr. Dalton's Theory of the constitution of mixed Gases. Phil Mag 19:193–196
14. Henry W (1803) Experiments on the quantity of gases absorbed by water, at different temperatures, and under different pressures. Phil Trans Royal Soc Lond 93:29–43 (with appendix 274–276)
15. Dalton J (1805) Water and other liquids. Man Mem 1:271–287
16. Farrar WV, Farrar KR, Scott EL (1975) The Henrys of Manchester. Part 4: William Henry: Hydrocarbons and the gas industry: minor chemical papers. Ambix 22:186–204
17. Henry W (1805) Experiments on the gases obtained by the destructive distillation of wood, peat, pit-coal, wax, etc. with a view to the theory of their combustion, when employed as sources of artificial light; and including observations on hydro-carburets in general, and the carbonic oxide. Nicholson's J 11:65–74

18. Henry W, Thomson T (1808) Experiments on the fire-damp of coal mines. Nicholson's J 19:149–153
19. Henry W (1808) Description of an apparatus for the analysis of the compound inflammable gases by slow combustion; with experiments on the gas form coal, explaining its application. Phil Trans Royal Soc Lond 98:282–303
20. Henry W (1819) Experiments on the gas from coal, chiefly with a view to its practical application. Man Mem 3:391–429
21. Henry W (1821) On the aeriform compounds of Charcoal and Hydrogen; with an account of some additional experiments on the gases from oil and from coal. Phil Trans Royal Soc Lond. 111:136–161
22. Henry W (1799) An Epitome of Chemistry: In Three Parts. J. Johnson, London

Chapter 12
Joseph William Mellor (1869–1938)

Joseph William (Joe) Mellor (Fig. 12.1) was born on July 9, 1869, in the small town of Lindley some two miles from the centre of Huddersfield in Yorkshire, the third child of Job and his wife, Emma (née Smith). His elder brother Frank had died at two years of age in March 1869 before Joseph was born leaving his sister, Ada, his only sibling at birth. His parents subsequently had a further six children of whom younger brother Oliver (1874) and sister Amy (1875) each survived only a matter of months. His other sisters Ada, Agnes and Gertrude moved to New Zealand with Joseph and their parents when they emigrated from the UK in 1879. The youngest child, Nellie, was born on the voyage and given Hurunui as her middle name, after the sailing ship in which they had passage. A youngest brother, Alfred, was born in Dunedin in 1883 [1].

The family arrived in Lyttleton, New Zealand, on December 6, 1879, after a 114-day voyage [2]. Job Mellor found work in the mill of the Kaiapoi Woollen Manufacturing Company and the family settled in the north Canterbury town. The then newly established company took on many workers from the north of England because of their skill in spinning and weaving, but the Mellor family's time there was a mere two years as Job found better prospects in Dunedin. He settled his family there permanently in 1881 when he gained work in the Ross and Glendining Roslyn Worsted and Woollen Mills [3, 4]. The firm had begun operations in 1879 in the Kaikorai valley as one of a number of factories and Job Mellor settled and built his house in the valley. The children went to the local Linden (the original name for the area) School, which became the (New) Kaikorai School in 1884. As noted above, the last of Job and Emma Mellor's children, Alfred, was born in 1883 in Kaikorai, but little is known of him. It appears that he was a jeweller prior to serving in WWI in 1918 as a Private in the 34th Reinforcements Otago Infantry Regiment, D Company, but his life span is unknown.

A previous version of this chapter was published as Halton B (2014) Joseph William Mellor, CBE, FRS (1869–1938). Chem N Z 78:85–89.

 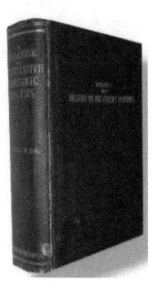

Fig. 12.1 Joseph William Mellor (courtesy of Special Collections, University of Otago Library) and the cover of *Quantitative Inorganic Chemistry* (1913)

The Mellor family was not wealthy and Joe left the Kaikorai School at age thirteen in 1882 to begin work as free secondary education did not exist at that time. Initially, he was a handy-boy for Mr. H. S. Fish, the then MP for Dunedin South and former (1870–73) and subsequent (1893–95) Otago City Mayor. From there Joe went to Simon Brothers boot shop and then to McKinley's boot factory, finally spending several years in the boot factory of Sargood, Son and Ewen, which employed upwards of 250 workers. As recorded in the Railways Magazine [3] (and elsewhere), the young Mellor was a 'boot clicker', the person who cuts the uppers for shoes or boots from a sheet of leather and named as such after the sound that the machine made. However, A.H. Reed in his summary of Mellor [5] cites first-hand evidence that Joe was a finisher, rubbing down soles to provide a good appearance and waterproofing, blackening and waxing the boots.

Mellor's foreman in the factory recollected him as being studious, pondering over mysterious books during the lunch hour and whenever the factory was at standstill [3–5]. Joe Mellor took to self-education, even building himself a small 1800 × 1800 mm corrugated iron shed in the garden at home for the purpose. The shed became his evening retreat, heated by a brick put in the kitchen oven over the evening meal by his mother and then wrapped in a flannel to keep his feet warm; it was lit by a small kerosene lamp. Here Joe carried out those experiments that his limited savings would allow and he read much. He borrowed books from wherever he could and then copied out the contents in longhand. His parents had much hope and expectation for their studious son. In his early teens, Joe confided to Arthur Ellis (a life-long friend who married his sister Agnes on Christmas Day in 1894) that he was determined to become the foremost chemist of his generation. Perhaps this enthusiasm came from his father's interest in anything scientific.

In any event, Joe's endeavours gained the attention of Mr. Thompson, the science master at Otago Boys' High School who was a founder, director and teacher at the Dunedin Technical School that opened in 1889. The twenty-year old Mellor took immediate advantage of the evening classes offered and matriculated in 1892, gaining the respect of Thompson to the extent that a scholarship to the University of Otago was arranged; Thompson even negotiated the time off work from Sargood's so that Mellor could attend lectures. As a part-time student it took Mellor until 1897 to graduate with his B.Sc. degree. Under the university's inaugural Professor of Chemistry, J. G. Black, Joe Mellor was awarded the Senior Scholarship in chemistry that year and he proceeded directly to Honours from which he graduated first-class and was offered the 1851 Scholarship in Science in 1898. This proved to be not as straightforward as might be imagined. The examination papers were set by specialists in England and the scripts returned there for marking. However, the steamship carrying them sank off Cape Horn and the candidates were re-examined [5]. Once offered, Joe accepted the 1851 electing to study chemistry at Victoria University (Owens College) in Manchester. Since the funds from the scholarship did not become available until later in 1899, the young graduate was employed as a lecturer in Natural Sciences at Canterbury (later Lincoln) Agricultural College until the middle of the year. On June 25 in Dunedin, Joseph William Mellor married Emma Cranwell Bakes, a gracious young lady from Lincolnshire, England, who had been brought up in Auckland and lived with her mother in Stanley Street, Mornington, Dunedin. She was a music teacher and the organist at the Mornington Methodist Church. The newly married couple sailed from Port Chalmers for England a short time later and Joe took up his scholarship, never to return to New Zealand. The Mellors had no children and Emma remained her husband's constant companion, assisting him throughout his busy and hectic life. A. H. Reed records that Mellor was her 'Joe' and Emma his 'boss' [5].

Owens College in Manchester was the first constituent part of the federal Victoria University, England's first civic university comprising of Owens and colleges in Leeds, Liverpool and Birmingham. Once there, Mellor came to experience the subsequently noted organic chemist, W. H. Perkin Jr. and started work with him. His exploits into organic chemistry led to a 1901 publication 'Some α-alkyl substitution products of glutaric, adipic and pimelic acids' [6], which showed the chain extension of ethyl 4-chlorobutanoate [$ClCH_2(CH_2)_2CO_2Et$] with the sodium salt of diethyl malonate [$NaCH(CO_2Et)_2$] to be just as effective and more judicious with 1-chloro-4-cyanobutane [$ClCH_2(CH_2)_2CH_2CN$].

Despite this, it was physical chemist H. B. Dixon who had the greater influence on Joe and it was he who set him on his subsequent career [7, 8] from research into the combination of hydrogen and chlorine that ran to a series of seven papers in the various Chemical Society publications (Proceedings, Transactions and Journal) [9–13]. In essence, Mellor repeated the earlier 19th century studies of Draper, Bunsen and Roscoe, and Pringsheim, showing the first two works to be correct while that of Pringsheim, who had suggested Cl_2O as the reaction intermediate, was not; Mellor proposed an alternative mechanism.

In addition to his research studies, Mellor began his 'second' career, namely that of an author, and a noted one at that. It was in 1901 that he had the idea of writing a book to make it easier for students to grasp concepts of the mathematics that one of his contemporaries in particular had had difficulties with when trying to follow chemical developments. Dixon encouraged him, the outcome of which was *Higher Mathematics for Students of Chemistry and Physics* [14] that ran to four editions and was reprinted in 1926. Prior to the completion of Mellor's Manchester studies, Professor Black at Otago University retired and it was suggested to him that Mellor be appointed as his successor; Black decreed that "he would be wasted here" envisaging a distinguished career for the man in England [3, 4]. Thus, with his doctoral studies completed and an Otago University [15, 16] (not Owens College [17]) D.Sc. degree, Dr. Mellor applied in 1902 for and accepted appointment as Science Master at Newcastle High School (now Newcastle-under-Lyme School). It seems that his application for this position was made on the mistaken assumption that he was applying for a post in Newcastle-upon-Tyne in northeastern England [16]. Thus, one may regard Joe Mellor's subsequent major contribution to the pottery industry as arising by chance since it was as a Staffordshire schoolmaster that he became fascinated by clay technology and pottery manufacture. It was during this time that he wrote his second book, *Chemical Statistics and Dynamics*, that appeared in 1905 [18].

In 1900, the North Staffordshire Ceramic Society was formed to foster discussion on matters relating to the clay working industries [16]. Past and current students of local pottery classes were invited to attend the first meeting of the society. From its small beginnings it evolved and expanded into the English (then British) Ceramics Society that is now a part of the Institute of Materials, Minerals and Mining. Dr. Mellor became associated with the society in about 1903, became a member in January 1905, and within six months was appointed vice-president and secretary, holding the latter post until his death in 1938; it is perhaps surprising that he never held the presidency [16]. The six pottery towns of Tunstall, Burslem, Hanley, Stoke, Fenton, and Longton (amalgamated in 1910 as Stoke-on-Trent) put pottery classes on a firm foundation by starting a pottery school in Tunstall in 1904 to which Dr. Mellor was appointed lecturer and head in pottery manufacture. The school provided technical and scientific training for managers in the pottery industry but the early years of the school were not easy. By 1907 the school had moved to temporary buildings in Stoke alongside those where mining classes had commenced the previous year [19]. Despite the rather meagre facilities, Mellor turned down the offer of the Chair of Chemistry at Sydney University in 1908, instead staying on in Stoke [20]. In 1910, after amalgamation of the six towns, land in Stoke was set aside for an educational institution. Largely due to the efforts of Dr. Mellor, plans for the building were drawn up and it opened in 1913 as the Central School of Science and Technology. It comprised large chemistry and physics laboratories, a large pottery laboratory, an analysis room, a grinding room, classrooms and lecture rooms, and other accommodation, while also providing for mining classes [21]. The old Pottery Department became the Ceramics Department to which Joe Mellor was appointed head where he established courses leading to degree level qualifications and organised research projects to help manufacturers produce better quality ware. As a teacher Mellor was popular with his students as his own early struggles gave him a deep insight into the difficulties of the evening-class student. Shortly before his death he is credited with saying "I think I spent some of the happiest years of my life with those early students" [22].

The early contributions that Mellor made to the ceramics industry were a 1904 paper in Transaction of the English Ceramics Society [6] and a lecture to the North Staffordshire Ceramics Society on Saturday April 15, 1905, entitled Crystalline glazes, which was accompanied by a note with J. Rodgers on pink glazes [16]. In all, Mellor contributed over 100 papers to the British (originally English) Ceramic Transactions ranging widely over the field. These included studies on the constitution of clay and the crazing, peeling and durability of glazes of which he was especially interested. However, his studies extended to the action of heat on refractory materials, the specific heat of firebricks at high temperature and the fine grinding of ceramic materials. In presenting his work [23] on the clay molecule (kaolin, $Al_2O_3 \cdot 2SiO_2 \cdot 2H_2O$), Mellor apologised to the English Ceramic Society for its purely chemical nature going on to say [6]:

> But clay is the lifeblood of pottery, and it is difficult to live with clay day after day without trying to form some idea of its nature and character. Today we may not know enough to see the practical bearing of this work but who dare predict what we shall see tomorrow.

In presenting this work he acknowledged proofreading by the noted English potter Bernard Moore and with whom Mellor had a life-long friendship. Mellor's early studies form the basis of understanding of clay firing, beyond which he contributed further to industry by providing his third text *The Crystallisation of Iron and Steel. An Introduction to the Study of Metallography* [24] in 1905. This book was based on the six lectures he delivered to the engineering students at the Staffordshire County technical classes in November and December of 1904.

Despite his active involvement in the creation of the Central School of Science and Technology, Dr. Mellor continued to perform research, write, and continue his teaching duties. In 1912 he published [25] his book *Modern Inorganic Chemistry* that went to eight editions (154,000 copies during his lifetime [20]), the last being prepared in 1939 by Dr. G. D. Parkes a year after Mellor's death. It ran to 871 pages in its initial edition and became the standard inorganic textbook throughout the English-speaking world and beyond. It was reprinted until 1951 and was abridged to provide three independent titles: *Introduction to Modern Inorganic Chemistry* (1914) [26], *Intermediate Inorganic Chemistry* (1930) [27] and *Elementary Inorganic Chemistry* (1930) [28]. Following the initial 1912 release came *A treatise on quantitative inorganic analysis, with special reference to the analysis of clays, silicates, and related materials* in 1913 and this became a standard work on silicate analysis, also until well after his death [29].

It was Mellor who recognised that almost every industry employing high temperature manufacturing processes in the early 20th century was limited in scope and efficiency by the durability of the refractory materials employed. He linked difficulties in the ceramics industry with those in firebrick manufacture. Indeed, it was largely as a result of his effort that the Refractories Committee of the Institution of Gas Engineers was formed in 1909, which instigated research into manufacture, properties and durability of refractory materials. Following from this came the 1920 British Refractories Research Association (BRRA) to which Dr. Joseph Mellor was appointed the first research director, a position he held until his retirement in 1937. That year the pottery industry was required by the Import Duties Advisory Committee to create a research association and the British Pottery Research Association was formed. This and the BRRA merged in 1948 as the British Ceramic Research Association and a new building was opened in 1951 in Penkull, a dormitory suburb of Stoke-on-Trent [30].

At the outbreak of war in 1914, the British steel and pottery industries were faced with a major crisis. The steel industry was desperately short of suitable refractory linings for the existing furnaces and needed more fire bricks to build new ones to increase output for the shipbuilding, engineering and munitions industries; the pottery industry needed temperature-indicating cones in order to continue. Importation of the traditional magnesite (calcined $MgCO_3$) bricks from Austria was banned as were the Seger cones from Germany. Liverpool University geologists showed that the UK had sufficient raw materials to manufacture furnace linings and fire bricks except for magnesite. Greece was to become supplier but the composition of Greek magnesite bricks differed in composition from the Austrian ones and required additional working. Aware of the problems the country was facing, and unable to join the armed forces because of ill health, Joseph Mellor approached government with an offer to use his expertise in the design of a new brick for the steel industry [31]. Working with his Staffordshire students in the Ceramics Department he developed a successful furnace lining that allowed the steel industry to maintain its production without even one day's interruption.

This work of Mellor appears to have been linked with that which followed an emergency meeting of the School's Governors that led to his services being offered to government to manufacture the vital pyrometric (Seger) cones (Fig. 12.2) commercially at the School [32]. These cones are triangular pyramids used to determine the internal furnace or kiln temperature and show when a material had been fired from placing inside a set of three or four cones and recording their bend at different temperatures. Working in a dedicated laboratory at the Central School of Science and Technology, Mellor developed a cone that became the Staffordshire Cone, stamped with the Stafford knot, the traditional symbol of the English county of Staffordshire and of its county town. A workman was employed and paid 25 shillings ($NZ 2.50) a week to make the cones under Mellor's supervision. The venture became so successful that Mellor's salary was increased to GBP 100 p.a. and two boys were employed from early 1915 to assist with the work and paid five shillings a week each (GBP 13 p.a.). Late that year an order for 200,000 cones was received. For the contributions that he made towards the war effort, Joseph was offered a peerage but he turned it down saying that he had given of his scientific knowledge freely to help his country because ill health prevented him joining the army and fighting in France.

Throughout the remainder of his career as a teacher and researcher in the pottery industry Joseph Mellor maintained his position at the forefront of development and

12 Joseph William Mellor (1869–1938)

Fig. 12.2 Left: Four pyrometric Segerkege (Seger) cones by Tinux (from Wikipedia); right: The Staffordshire knot (badge) from *English Heraldry*, by Charles Boutell (from Wikmedia Commons)

the British ceramics industry owes much of its 20th century reputation to the efforts of this man. As a major interest he researched glazes although his significant publications in the area did not appear until late in his career. Thus, his noted *The Crazing and Peeling of Glazes and The Durability of Pottery Frits, Glazes and Enamels in Service* appeared only in 1935 after his retirement from the Pottery Department at age 65 the previous year. However, Mellor was prodigious in his writing and from the early 1920s his attention focused more closely on what became his monumental *Comprehensive Treatise on Theoretical and Inorganic Chemistry* that was published in 16 volumes between 1927 and 1937 by Longman, Green and Co. It started with the information collected for his 1912 Inorganic Chemistry text. Mellor collected all the information, wrote out every word of the every volume including all the citations given and had but one secretary with a standard manual typewriter to produce copy for the publisher. This *opus magnum* provided what has undoubtedly been the major reference work in the field. Each of the 16 volumes received much praise and supplements were produced under an editorial board until 1980. For a 21st century chemist to contemplate the time and effort required for Mellor to produce some 15,320 printed pages of detailed science from longhand written script with no computer is incomprehensible. It is not surprising therefore, that Joseph W. Mellor is best remembered among the chemical fraternity as an academic author par excellence. But Joseph W. Mellor made equal contributions to academic chemistry from his writings and to applied chemistry from his work with pottery materials.

Mellor had yet another side—that of the humourist. Although he had no children of his own he maintained contact with his nephews and nieces in New Zealand writing letters and illustrating them with cartoons. These culminated in a book published for the Ceramics Society by Longman Green in 1934 entitled as shown below (Fig. 12.3) and illustrated by the cartoon [32]. In addition to this Mellor was a part of the 1929 Ceramic Society delegation to the US on the Cunard liner, R. M. S. Laconia. For this, Joe produced a 36-page logbook of the voyage, which contains the menus of the eight-day voyage each adorned with two full-page cartoons. It has been described

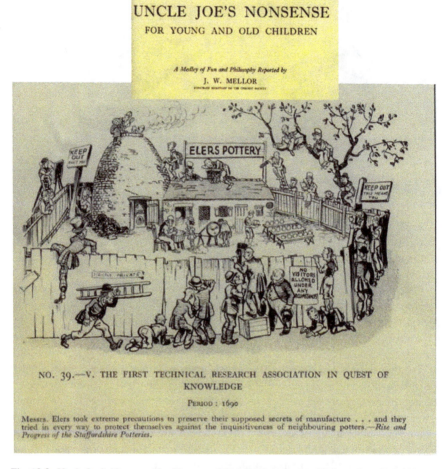

Fig. 12.3 Uncle Joe's Nonsense for Young and Old Children by Joe Mellor, taken from [32] and in Portraits of Potteries People, www.thepotteries.org/potraits/001/htm (courtesy of Steve Birks) (accessed 7 Oct 2013)

appropriately by Alexander Silverman who describes him not only as a wit but an artist par excellence [33].

For his work Joseph Mellor became the second person connected with the ceramics industry elected Fellow of the Royal Society in May 1927. The first was Joshia Wedgewood (January 16, 1783), the noted potter who invented the pyrometer to measure internal furnace temperatures in 1782. Upon retirement from the BRRA in 1937 Mellor received the CBE in the Coronation Honours and several authors have suggested that this was hardly fitting for someone who had made so many contributions to his country. He died in London on May 28, 1938 and, by 1960, the Mellor Building was constructed on the North Staffordshire Technical College site, now Staffordshire University.

References

1. See: http://freepages.genealogy.rootsweb.ancestry.com/~kiwiadams/10300.html. Accessed 4 Sept 2013
2. Nelson R (2001) New Zealand migrant shipping (1876–1885). http://mem-bers.iinet.net.au/~perthdps/shipping/mig-nz3.htm#nz3. Accessed 4 Sept 2013
3. J S (1938) Dr. J. W. Mellor, F.R.S., A world chemist and humourist from Otago University. NZ Railways Mag 13(6):9–17. http://nzetc.victoria.ac.nz//tm/scholarly/tei-Gov13_06Rail-t1-body-d3.html. Accessed 31 Aug 2013
4. Jenkinson SH (1940) 10-Mellor. In: New Zealanders and Science. Department of Internal Affairs, Wellington, pp 104–116. http://nzetc.victoria.ac.nz/tm/scholarly/tei-JenNewZ-t1-body-d10.html. Accessed 1 Oct 2013
5. Reed AH (1957) Joseph William Mellor. Dunedin Public Library Publication No. 2, A.H & A.W Reed, Wellington (reprinted by the Otago Daily Times, January 1957)
6. Green AT (1939) Joseph William Mellor, 1869–1938. Obit Not Fellows R Soc 7:573–576
7. Habashi F (1990) Joseph William Mellor. Bull Hist Chem 7:13–16
8. Mellor JW (1901) Some α-alkyl substitution products of glutaric, adipic, and pimelic acids. J Chem Soc Trans 79:126–134
9. Mellor JW (1901) On the union of hydrogen and chlorine. Parts I to III. J Chem Soc Trans 79:216–238
10. Mellor JW, Anderson WR (1904) The union of hydrogen and chlorine. Part IV. The Draper effect. J Chem Soc Trans 81:414–418
11. Mellor JW, Russell EJ (1902) The preparation of pure chlorine and its behaviour towards hydrogen. J Chem Soc Trans 81:1272–1280
12. Mellor JW (1902) The union of hydrogen and chlorine. V. The action of light on chlorine gas. J Chem Soc Trans 81:1280–1292
13. Mellor JW, Russell EJ (1902) The preparation of pure chlorine and its behaviour towards hydrogen. J Chem Soc Trans 81:1292–1301
14. Mellor JW (1902) Higher mathematics for students of chemistry and physics. Longmans, Green & Co., London
15. University of New Zealand (1963) Alphabetical roll of graduates 1870–1961. Whitcombe & Tombs
16. Booth R (2000) Mellor—the legend and the legacy, 44th Mellor Memorial Lecture. Br Ceram Trans 99:165–174
17. Habashi F. Mellor Joseph William. Oxford Dictionary of National Biography. www.oxforddnb.com/view/article/56145?docPos=1. Accessed 23 Sept 2013
18. Mellor JW (1904) Chemical Statistics and Dynamics. Longmans, Green & Co., London
19. Joseph Mellor—the man who turned down a peerage. www.northstaffordshire.co.uk/?p=3554. Accessed 25 Sept 2013
20. Wilkins CJ (1966) Mellor, Joseph William, C.B.E., F.R.S. Te Ara encyclopedia of New Zealand. www.teara.govt.nz/en/1966/mellor-joseph-william-cbe-frs. Accessed 12 July 2013
21. Birks S, Neville Malkin's "Grand Tour" of the Potteries. www.thepotteries.org/tour/084.htm. Accessed 24 September 2013
22. At G (1938) Dr, J. W. Mellor, C.B.E., F.R.S. Nature 142:281–282
23. Mellor JW (1911) The chemical constitution of the kaolinite molecule. Trans Eng Ceram Soc 10:1911
24. Mellor JW (1905) The crystallisation of iron and steel. An introduction to the study of metallography. Longmans, Green & Co., London
25. Mellor JW (1912) Modern inorganic chemistry. Longmans, Green & Co., London
26. Mellor JW (1914) Introduction to modern inorganic chemistry. Longmans, Green & Co., London
27. Mellor JW (1930) Intermediate inorganic chemistry. Longmans, Green & Co., London
28. Mellor JW (1930) Elementary inorganic chemistry. Longmans, Green & Co., London

29. Mellor JW (1913) A treatise on quantitative inorganic analysis, with special reference to the analysis of clays, silicates, and related materials. C. Griffin Co., London
30. Birks S, Joseph Mellor. Portraits of Potteries People. http://www.thepotteries.org/portraits/001.htm. Accessed 8 Aug 2018
31. Cooper B, Martin D, Central School of Science and Technology, The Phoenix Trust 2010. http://www.northstaffordshire.co.uk/?tag=central-school-of-science-and-technology. Accessed 4 Sept 2013
32. Mellor JW (1934) Uncle Joe's nonsense for young and old children. Longmans, Green & Co., London
33. Silverman A (1952) Mellor's nonsense. J Chem Educ 29:187–188

Chapter 13
John Mercer (1791–1866)

John Mercer (Fig. 13.1) was born on February 21, 1791 in the town of Great Harwood in Lancashire, England, the second son of a Lancashire cotton spinner [1]. At the time of his birth his father ran a cottage industry spinning mill by the side of Dean Brook, the stream that feeds the Clayton-le-Moors and Great Harwood reservoir. The advent of machinery and the formation of large cotton mills led his father to alter tack and he leased '*The Stoops Farm*' to the east of the small township off the road to Whalley. In 1799, there was a major crop failure in the district and the virulent epidemic that followed claimed many lives, including that of John's father; he died on August, 7, 1802 when John was eleven years old. By then John was in his second year of work as a bobbin winder, subsequently to become a hand-weaver. The family was very poor. John's mother, Betty, remarried in 1806, and he became half-brother to William, who was born that year.

John lived with various relatives over the years. However, when John was 10 years old Mr. Blenkinsop, a neighbour and pattern designer at the nearby Oakenshaw Printworks in Clayton-le-Moors (where cotton calico[1] cloth was dyed) started to teach John to read and write and introduced him to long division in mathematics before he moved away [2, 3]. John continued his education himself and gained a reputation as 'adept at figures' [2]; he also became a self-taught musician and played several instruments and formed a choir and a band [2]. Later, John Lightfoot, who was the Excise surveyor at the same calico dye works (each square yard of printed calico was levied with threepence excise duty) and known to visit Harwood frequently, befriended John and taught him higher mathematics, teaching him with his own sons who were calico dyers at the works [3, 4]. Mercer was a keen learner and soon became even more recognized for his aptitude with figures and the skill he had from his self-

---

[1] Calico is a plain-woven textile made from unbleached, and often not fully processed, cotton.

A previous version of this chapter was published as Halton B (2013) John Mercer FRS, FCS, MPhS. Chem N Z 77:22–26 and Halton B (2013) John Mercer Part II. The industrialist, the chemist and the man. Chem N Z 77:52–55.

Fig. 13.1 John Mercer (Portrait von John Mercer dem Erfinder der Merzerisation Mercer starb 1866, daher ist das Bild gemeinfrei Quelle) from Lancashire Pioneers via Wikipedia

taught music. The wars of the era required John to join the militia, something that he found uncongenial as he was inept and too clumsy. He was put in the "awkward squad" that led to his nickname of *Awkward John*; subsequently, he was transferred to the band and music [2].

On one occasion, when visiting his mother he saw his half-brother, William, seated on her knee and wearing an orange dress. That single vision changed his life forever as he decided that he should become a dyer. As quoted in the book by his nephew Parnell [2, 3], John Mercer was "all on fire to learn dyeing", but he had had no instruction in the subject, no books, nor the means to obtain them. However, he found that the dyers of the area bought their supplies from a druggist in Blackburn, a larger town some eight kilometres away. He went there and asked for dye-stuffs, but had no idea what it was he needed.

The druggist gave him the names of the common materials then in use: peach wood (*Caesalpinia echinata*) from a tropical tree with a prickly trunk and Brazil wood (another form of *Caesalpinia echinata*, which is a dense, orange-red heart-wood that takes a high shine and is the premier wood used for making bows for stringed instruments), which yield the red pigment Brazilin known as Natural Red 24 (**1**, Fig. 13.2); alum [potassium alum, the potassium double sulfate of aluminium often designated, $KAl(SO_4)_2 \cdot 12(H_2O)$ but correctly $Al_2(SO_4)_3 \cdot K_2SO_4 \cdot 24H_2O$], which when added to water clumps the negatively charged colloidal particles together into flocs; copperas [iron(II) sulfate, $FeSO_4 \cdot 7H_2O$], which dissolves in water to give the blue-green $[Fe(H_2O)_6]^{2+}$) complex; logwood (*Haematoxylum campechianum*) from which hematoxylin or Natural Black 1 (**2**) is extracted, and the oxidation product of which is hematein (**4**); and quercitron, the yellow natural dye (**3**) obtained from the bark of the Eastern Black Oak (Quercus velutina), a forest tree indigenous to North America [5].

Mercer checked his money and found that he could afford to pay threepence for each dyestuff. He purchased the materials available. He was very fortunate in being allowed a suitable place to begin his experiments where he had the necessary equipment for his trials, which were carried out as rule-of-thumb experiments. Thus,

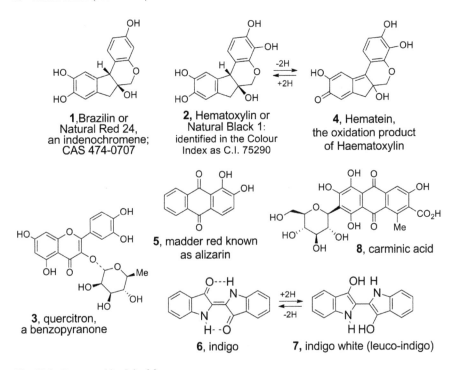

Fig. 13.2 Dyes used by John Mercer

it was by close observation and the maintenance of accurate records that he acquired considerable knowledge of the properties of the dyestuffs and the way in which to provide the colours used in dyeing then in vogue. Some mineral dyes such as Prussian blue, manganese bronze, chrome yellow, antimony orange, and iron buff pigments were fixed to cotton with the use of egg albumen or blood, but wheat gluten or milk lactarine (casein) were also used. Heat and acid were also needed to make them colourfast. Brazilin had been used since at least the Middle Ages to dye fabric, and in the formulation of inks as well. The specific color produced depends on the manner of preparation: in an acidic solution it is yellow, but in an alkaline preparation it is red. In contrast, alum was added to the water so that the negatively charged colloidal particles clump together into flocs.

By the age of 16 years Mercer was a hand-loom weaver and, following his experiments with the Blackburn dyes, he vowed to become a dyer and in this he was successful. He formed a business with a partner using the remnants from the Great Harwood loom weavers and gained success to the extent that his experiments with dyestuffs attracted the attention of the Fort brothers, the owners of the Oakenshaw Printworks. In September 1809, when he was 18 years of age, he was offered an apprenticeship in the colour-shop at the works [2, 3]. However, the old foreman there felt threatened by John and offered him no useful information, giving him instead duties more becoming an unskilled labourer [2]; the foremen of colour-shops tended to keep their dyer's art a secret and their mixes empirical.

By 1810, Napoleon's decrees on trade (1793–1810) had reached the point where all printed calico and other British manufactured goods arriving in France were burnt and no more were permitted to be imported. The impact of this at the Oakenshaw site was so severe that the owners offered for surrender their apprentice indentures to those who chose to leave. So it was that after ten months of apprenticeship John Mercer accepted the offer and returned to the hand loom.

At about this time John gained lodging with the Wolstenholme family and stayed with them for some time. In 1813 he converted to "the truth of Christian religion", and had resumed work as a dyer as well as weaving and again he was successful. However, it was not until 1814 that he was able to gain more chemical knowledge. By then he had become engaged to Mary Wolstenholme, some six years his senior, and described as "a very superior woman" [2]. A licence to marry could not be obtained in Great Harwood but required attendance at the office in Blackburn and, whilst there, John visited a second-hand bookstall on the market where, it appears, he devoted more attention to securing *The Chemical Pocket Book* or *Memoranda Chemica: arranged in a Compendium of Chemistry* (James Parkinson, 3rd edn., 1803) than to gaining the marriage licence. This was not his first chemistry book as John Lightfoot had presented him [2] with a copy of the 1787 *The Table of New Nomenclature* proposed by De Morveau, Lavoisier, Bertholet, and De Fourcroy. However, on its own the Table was of little help. The Parkinson book, however, opened up a new world, especially when coupled with the Table. From his earliest experiments, John adopted the view that that is was only through a thorough knowledge of the properties of the dyeing materials and their behaviour under a variety of conditions that the operation of the dyer could be performed intelligently. The books convinced him that all this knowledge depended upon chemical science and that it was on chemistry that the extension of his art rested.

John Mercer married Mary Wolstenholme on 17 April, 1814 and their first child, Mary Mercer, was born on 27 November that year, but survived a mere 17 days. It was not until 1817 that his second child, Mary Clayton Mercer, was born; she was the first of a further five children (three girls and two boys), none of whom married. John continued in his chemical quest and his first major discovery came in 1817. It related to the orange coloured clothes that he had seen his step-brother wearing.

Given the short account of 'the sulphide of antimony' in the chemical pocket book, Mercer performed a series of experiments and then tested his resultant formulations on calicos available. The results gave rise to his "antimony range" of dyed calico. He found that the alkaline sulfantimonates (salts of the hypothetical sulfantimonic acid, H_3SbS_4) provided an excellent medium to give a bright orange colour on calico—he fixed antimony sulfide (Sb_2S) to produce orange calico prints, something that had previously been unattainable. At that time, orange colours came from mixes of quercitron yellow (**3**, Fig. 13.2) and madder red (**5**), but Mercer's antimony orange more appropriately supplied the need. Moreover, it was capable of combination and interspersion to give a good variety of styles unlike the madder-quercitron combination.

13 John Mercer (1791–1866)

Mr. Lightfoot advised John to make his discovery known and available to Hargreaves, Dugdale and Co., the proprietors of the Broad Oak works in Accrington, which he did. On his way to Accrington to provide the necessary instruction to Mr. Hargreaves' dyers he happened to meet Mr. John Fort, his former employer at the Oakenshaw works. Fort had heard of the Mercer "antimony orange range" and took the opportunity to offer him the job of experimental chemist at his works at an initial salary of 30s ($NZ3) per week [2], something that Mr. Hargreaves had omitted to do. This led to Mercer being reemployed by the Fort brothers, by then 1818, when formally he became a chemist in the colour shop. His successes were such that he was offered and accepted a partnership in the company in 1825, and he remained associated with the firm until its dissolution in 1848, at which time he elected to retire, by then a rich and famous man.

At this time all dyestuffs were supplied in crude state, with nothing available from synthetic preparation [5]. It was, therefore, important to purify the dyes to avoid overspill (the dye spreading beyond the desired part) in the dyeing process, and chemists were employed at the various dye works for this and to improve the process.

From the time of his employment (previously, an apprentice until 1810), Mercer's skill and energy provided many inventions of new styles and new colours in the dye industry and, importantly, they led to new branches of industrial chemistry, as it was then known [2, 3]. It was his "antimony range" of dyes that led to his (re)employment, and this he developed further to give browns of differing hues on calico[2] with copper(II) and lead(II) compounds [5]. His work with chromium compounds is reputed [2, 3] to have created the industrial manufacture of bichrome (potassium dichromate, $K_2Cr_2O_7$). At the time of his initial experiments with it, the cost was ten shillings and six-pence an ounce but by 1885, after his studies and with industrial manufacture becoming more common, its cost was less than sixpence per pound. He rediscovered and introduced into England a method of applying lead chromate ($PbCrO_4$) to cotton cloth in 1823. This yellow dye was then of great importance and formed within the fabric by treating it with lead acetate and then potassium chromate. That same year he introduced manganese bronze (a mixture of manganese oxides whose hue depends upon the degree of oxidation) formed from manganese salts, a dyestuff that went in and out of fashion in a cyclic manner almost every ten years. He also greatly improved the methods of printing indigo by using potassium ferricyanide [$K_3Fe(CN)_6$] and potash.

Of almost equal importance was his method of preparing mixed cotton and woollen fabrics that would subsequently accept dye with equal effectiveness. He found that some aluminium lakes [a lake is a pigment formed by precipitating a dye with an inert binder (the mordant), usually a metallic salt] from organic dyes were dissolved by ammonium oxalate [$(NH_4)_2C_2O_4$], and this led to the use of aluminous colour-precipitates in steam colour work. He also devised a new and cheaper method of preparing sodium stannate (Na_2SnO_3), a compound that was valuable to the calico dyer providing him with much monetary gain.

---

[2]See Footnote 1.

Of the numerous improvements made by Mercer, the use of alkaline arsenites, e.g., Na_3AsO_4, in the dunging operation was one of the more important [2]. Superfluous uncombined mordant that remained on the cloth from the first stages of the dyeing process, and any remaining thickening agent with which the mordant was printed in, had to be removed after ageing and prior to attaching the dye. Unless adequately removed, the mordant dissolved in the dye-bath and caused discolouration of the whites, a deterioration of the dyeing liquor, and the dye spread beyond the defined pattern. Traditionally, the removal was accomplished by passing the cloth through hot water in which cows' dung had been dissolved—hence the name. Mercer's discoveries led to far superior substitutes, the principal ones of which were sodium silicate and arsenate.

Once established, John Mercer was sought after by chemical manufacturers and dyers, and he is reputed to have given his services freely. His self-taught skills and knowledge had, from the earliest times, led him to make accurate records of his work to the extent that solution strengths and chemical quantities were recorded in (the then new) equivalents. He is regarded as one of the earliest workers in volumetric analysis. In 1827, he was able to value bleaching powder and bichrome using standard solutions. He speculated on the nature of white indigo **7**, the educed soluble form of the indigo, **6** (Fig. 13.2). When calico is submerged in a bath of white indigo and then removed, the white indigo quickly reacts with atmospheric oxygen and reverts to the insoluble, intensely colored indigo in the fabric. His views on what happened here were contrary to the accepted ideas but, subsequently, the redox process as we know it was proved correct [2].

From the time of his first experiments early in the nineteenth century, Mercer had been concerned with avoiding waste. In about 1825, he introduced a new way of recovering the colouring matter of cochineal (the aluminium salt of carminic acid, **8**). The traditional method of the time, involved digesting in hot water, but Mercer found that some 25% of the dye was left in the dregs. At ten shillings per pound of weight this was an unnecessary waste. By adding a neutral alkaline oxalate to the water, he found that 100% of the dye was extracted. Some 25 years later a French chemist visited the Oakenshaw works [2] to describe his method of cochineal extraction, but he had to concede that, although similar, the Mercer method required fewer steps and was superior.

In 1841, Lyon Playfair (the noted chemist who had been James Thompson's private assistant and then a Ph.D. graduate of Liebig in Giessen) was employed as a chemist at Thompson's Primrose Printworks in Clitheroe (a town some 12 km from Great Harwood). He recognized Mercer's abilities and the two men became friendly. They began to meet weekly in a Whalley pub mid-way between the towns, to discuss science and matters of the day affecting the printworks. These meetings attracted as many as ten like-minded individuals and it was at one of these gatherings that Mercer advanced his theory of catalysis, which he illustrated with many examples [2]. Subsequently, he read this theory to the 1842 Manchester meeting of the British Association for the Advancement of Science. Playfair extended these concepts and they were finally settled by Kekulé. Playfair's influence had Mercer extend his knowledge and by the time that he formally retired he was regularly using the Berichte and other journals of the era.

Further observations made by Mercer in 1843, and discussed at these Whalley meetings, led Playfair subsequently to discover the nitroprussides, e.g., $Na_2Fe(CN)_5NO$. Playfair was so enamoured with the work of John Mercer that he persuaded him to become a foundation member (in chemical manufacture) of The Chemical Society in 1842 (FCS). Later, a group of noted chemists of the day, James Thompson, Walter Crum and Lyon Playfair, on dining at the home of Professor Thomas Graham, concluded that Mercer should be nominated for election to the Royal Society. In a final letter to Mercer, Playfair stated that it "would be a great tribute to a man who has acquired knowledge of science without the aid of academies, and under every disadvantage" and it was this persuasion that had Mercer accede [2] He was elected a Fellow of The Royal Society (FRS) in 1852.

It was Mercer's discoveries of 1844 that gained him the fame and kudos he deserved, and a considerable fortune. He provided a formulation for red ink, which gave him the sum of £10,100, and then his experiments in treating cotton with sodium hydroxide, sulfuric acid and zinc chloride led to the mercerization process, which he patented in 1850. There appear to be few records of the red ink formulation, although his experiments were carried out in partnership with Robert Hargreaves at his Broad Oak factory, near Accrington. In contrast, the mercerization process is well known. What Mercer found was that any one of sodium hydroxide, sulfuric acid or zinc chloride caused individual cotton fibres to become thicker and shorter, giving the cloth much greater strength. It also became semi-transparent and better able to absorb dye. In his original process, the overall size of the fabric shrank, with the result that the process became popular only after H. A. Lowe improved it in 1890; the cotton was held during treatment to prevent it shrinking. The fibres then gained a lustrous appearance. This represents the modern form of the process with mercerized cotton now the norm.

Although discovered by Mercer in 1844, the fine details of the mercerization process are still not fully understood. Cellulose consists of a polymeric chain of D-glucose molecules depicted by structure **9** (Fig. 13.3). In the polymer, the alternate rings are rotated 180° with respect to one another and the chain is strengthened by H-bonding, as shown in **10**. The discourse by Speakman [6, 7] provides the essentials of what is known about the process of mercerization. Thus, the change in lustre of cotton fibres on treatment with concentrated NaOH is due to a change in the shape of the fibres. On the cotton plant the fibres are attached to the cotton seeds in a closed cotton pod. The fibres are straight with a circular cross-section and have a central cavity filled with a dilute aqueous solution of various salts and other small molecules, akin to thick-walled tubes. When the fibre reaches its final length, the pod opens, the fibres dry in the sun, flatten and twist and acquire a tortured shape. Electron microscopy and X-ray diffraction experiments have shown that the cell walls of the cotton fibres contain microfibrils, each composed of about 1000 fully extended cellulose molecules. Each of these molecules consists of some 6000 glucose residues. Along the microfibril, ordered regions that are crystalline alternate with shorter non-crystalline ones. Untreated cotton is some 60% crystalline and has all the cellulose

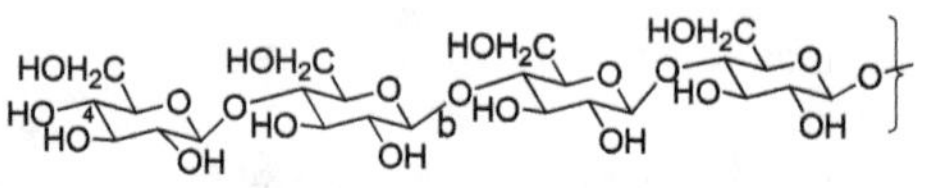

9, Cellulose: β-linked poly-D-glucose

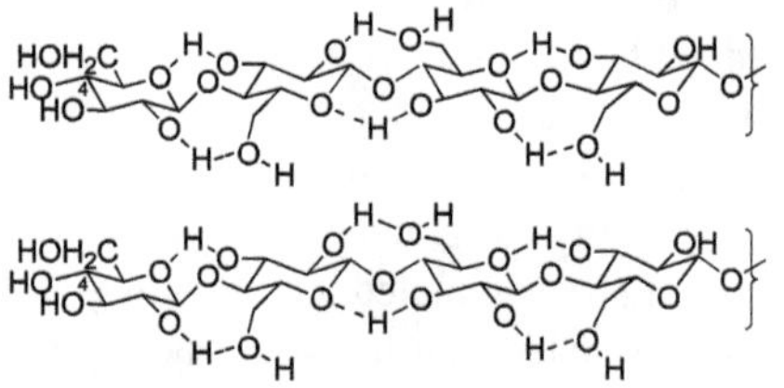

10, crystallized region of cellulose, showing H-bonding in the aligned parallel chains of the less thermodynamically favoured sheet of non-mercerized cotton

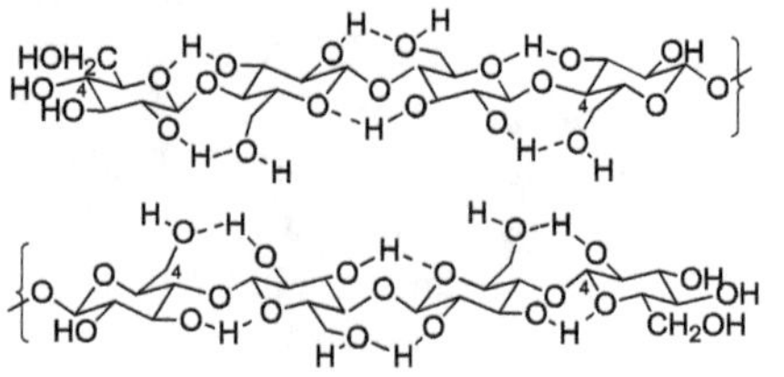

11, crystallized region of cellulose showing H-bonding in the parallel chains and their opposte orientation in the thermodynamically favoured sheet of mercerized cotton

Fig. 13.3 Cellulose and crystallised regions

chains running parallel to one another and is stabilized by H-bonding, as shown by **10** in Fig. 13.3. The chains are arranged in sheets as depicted in Fig. 13.4 for three cellulose molecules in two separate sheets (not to scale). The bonds shown in bold depict one sheet and the plain bonds the other. There is no H-bonding between sheets and the cotton is in its less thermodynamically stable form.

During mercerization, NaOH enters the central core of the fibril and breaks the H-bonding network and the fibres swell:

$$R-OH + NaOH \rightarrow R-ONa + H_2O$$

The chains can move with respect to the other. However, when the NaOH is washed out of the cloth the sodium salts revert to their normal hydroxyl nature and the H-bonding network is reestablished giving new microfibrils. The sheets are disrupted during the NaOH treatment and the regenerated microfibrils are formed in such a way as to generate the thermodynamically more stable form. The cellulose chains now run in opposite, antiparallel, directions depicted by **11** in Fig. 13.3 and the new sheet format of Fig. 13.4. Since the sheets do not align one with the other, the independent sheets carrying **9** do not lie in the same direction, and it appears that the antiparallel arrangement comes from new H-bonding between cellulose molecules of different sheets when the NaOH is removed. No single chain rotates through 180°.

An alternative explanation involves the stacking of the fibrils during the crystallization process. Here neighbouring parts of the molecule run in opposite directions in the zigzag plates and are antiparallel allowing for the reorganization [8]. Furthermore, stirring the polymer molecule during crystallization causes a portion of the long molecules to become extended, while the remainder crystallize to give plates at intervals along the rods. These rods are disc-shaped and give rise to a partly crystalline structure with wheel-like plates along its axis—such a structure is termed a synthetic polymer shish-kebab (a β-cylindrite structure). Mercerized cellulose has

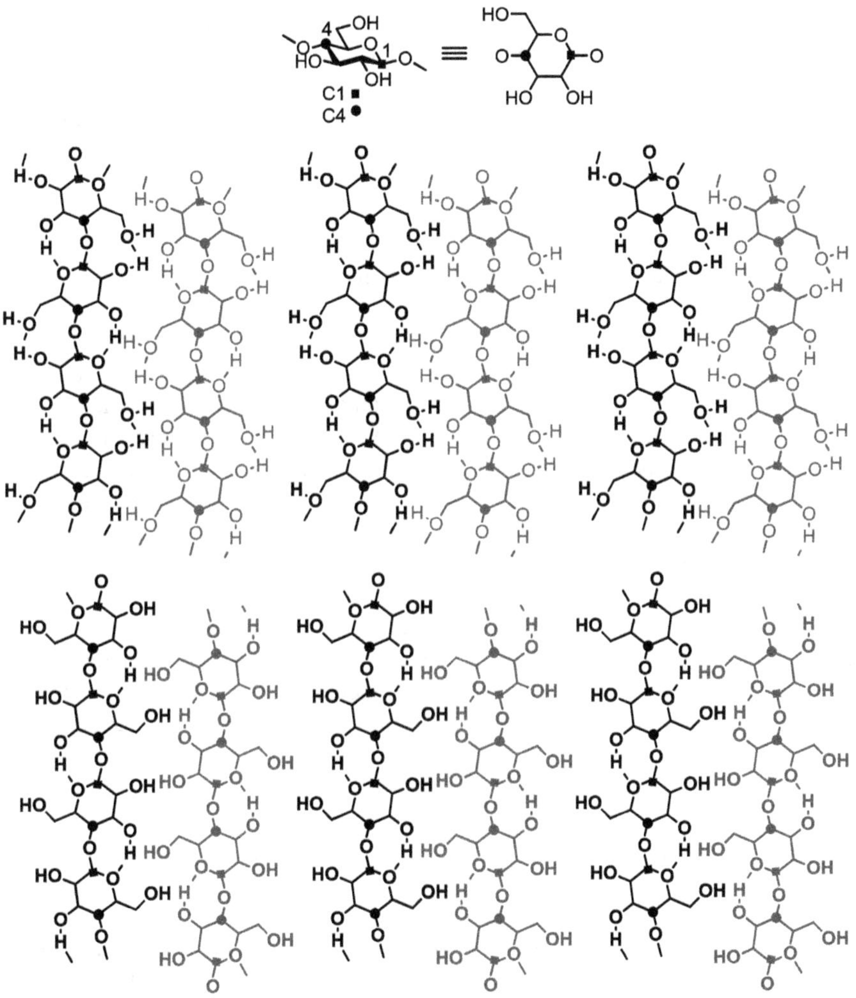

Fig. 13.4 Upper: two parallel sheets of cellulose unmercerized; lower: two parallel sheets of cellulose mercerized

been shown to have such shish-kebab components by electron microscopy [8]. The chains in the kebab parts of the cellulose shish-kebab molecule are folded in just the same way as from an unstirred solution and are, therefore, likely to be antiparallel; the chains in the shish (extended) part of the molecule are assumed also to be antiparallel. Although many fine details still remain to be established as illustrated by a recent review [9], Mercer's discovery of treating cotton cloth with sodium hydroxide to facilitate the dyeing process and imparting lustre means that it is entirely fitting that the commercial process immortalizes his name.

Mercer's immersion in colour led him to produce some of the earliest recorded colour photographs. In a letter to Lyon Playfair written in May 1857, he states that he had been experimenting for 2–3 years making photographic experiments, not as an artist but as a chemist. However, the earliest record of him observing the effects of sunlight on material dates back to the 1820s. His studies and 'amusements' led to him discovering a new way of measuring the power of solar rays and to the blue-print concepts. From 1854 he began producing colours on paper and cloth using the impact of light on various chemicals and obtained pictures "in a variety of pleasing tints" [2]. His studies led to him exhibiting a number of the photographs at the 1858 British Association meeting in Leeds. A professional photographer of the day said that he had never seen anything on paper as beautiful as the Prussian blue in softness and gradation that Mercer had obtained [2]. His specimens attracted much attention in Vienna and the photographic society there sought the process from him.

At the 1858 meeting of the British Association, Mercer described the reducing action of light on complex salts of iron to give a blue colour, the depth of which depended upon the light intensity, a discovery of his that dated to 1828. This 'blue-printing' proved of immeasurable value to the drawing offices. He intended to use blue-printing to measure the intensity of sunlight, an idea that was adopted by several metrological stations.

At the same meeting of the British Association, John Mercer advanced the concept of simple mathematical relations between the atomic weights of the elements. Unfortunately, this was generally ignored. Mendeleev's Period Table of 1869 and the subsequent periodic system followed [10], yet the extract from Parnell's book [2] shows that Mercer had a grasp of the octet and the periods (based on $O = 8$).

The principal partners of the Oakenshaw Printworks retired in 1848 and John Mercer elected to do the same, devoting himself more to experimental study. However, after his wife's death in 1859 his enthusiasm waned. He died on November 30, 1866 following complication from falling into the water of a reservoir (under construction) some two years earlier. It appears that he was an unselfish person with a furtive mind given to flights of scientific fancy with his friends. He was an experimentalist par excellence and had remarkable insight into chemistry. It has been said that had he devoted himself entirely to research, he would have been among the most distinguished chemists of the day [4]. Although he patented some of his inventions, he freely gave away many others to great advantage and considerable profit of the recipients. He became an honorary member of the Manchester Philosophical Society in 1849 and of the Glasgow Philosophical Society (MPhS) in 1860. He accepted a role equivalent to Justice of the Peace in 1861, but Parnell says that "the claims of justice to the community were not infrequently outweighed by his feeling of pity and compassion for the "offender". It is clear that John Mercer was an upstanding and outstanding member of the community.

References

1. Lancashire Pioneers. Lancashire Lantern. www.lancashirepioneers.com. Accessed Dec 2012
2. Parnell EA (1886) The life and labours of John Mercer, F.R.S, F.C.S., etc., the self-taught chemical philosopher. Longmans Green, London
3. Thorpe TE (1886) The life and labours of John Mercer, F.R.S. Nature 35:145–147
4. Hartog PJ, Brock WH. John Mercer (1791–1866). The Oxford Dictionary of National Biography. www.oxforddnb.com/view/article/18573?docPos=3. Accessed Dec 2012
5. Séquin-Frey M (1981) The chemistry of plant and animal dyes. J Chem Educ 58(4):301–305
6. Speakman P (1991) Shirts and shish-kebabs; John Mercer and mercerization. Biochem Educ 19(4):200–203
7. Langan P, Nishiyma Y, Chanzy H (1999) A revised structure and hydrogen-bonding system in cellulose II from a neutron fiber diffraction analysis. J Am Chem Soc 121:9940–9946
8. Chanzy HD, Roche EJ (1975) Fibrous mercerization of Valonia cellulose. J Polym Sci Polym Phys 13:1859–1862
9. Habibi Y, Lucia LA, Rojas OJ (2010) Cellulose nanocrystals: chemistry, self-assembly, and applications. Chem Rev 110:3479–3500
10. Mendelejeff D (1869) Ueber die Beziehungen der Eigenschaften zu den Atomgewichten der Elemente. Zeit Chem 12:405–406

Chapter 14
Ellen Swallow Richards (1842–1911)

Ellen Henrietta Swallow (Fig. 14.1) was born on December 3, 1842 on a farm near the village of Dunstable, in rural northern Massachusetts. She was the only child of well-educated but relatively poor New England parents, Peter Swallow and Fanny Gould Taylor [1–6]. Her father divided his time between teaching and farming having inherited half of his father's farm. Ellen received most of her early education from them at home. Her life was directed to being helpful and assisting others as expected in that puritanical region of the US. However, she focused on new ways of doing it with a passion for usefulness and a longing for pioneering. When she was 16 years old her father sold his farm and moved to a general store in nearby Westford so as to raise the money for his daughter's education. Ellen worked in the store to make her contribution to this. Her work serving behind the counter grew to include the bookkeeping and purchase of supplies.

Her father was her most ardent supporter in gaining a college education and a scientific training at a time when such things were almost unknown among women. She was enrolled at the then private Westford Academy, one of the oldest high schools in the US, graduating in 1862 having studied much Latin, some French and a little Maths [1]. She was to have started teaching after leaving the academy but a bad bout of measles and its aftermath prevented this and it was only in the following May that her teaching career began. In 1863, to enlarge his business, her father moved the family to nearby Littleton. Her mother had been and was frequently ill and so, by early 1865, Ellen was back at home looking after her, the house, and helping her father in the store. As Littleton was close enough to Worcester for Ellen to attend lectures she moved and lived there for the 1865–1866 winter. After returning home Ellen focused her energy on assisting in the store, and spent long periods looking after her ill mother, still determined to continue formal education despite a lack of money. She saved every penny despite the 1866–1868 years being ones of ill health and deep depression. Nonetheless, by the time she was almost 26 she had saved

---

A previous version of this chapter was published as Halton B (2016) Ellen Swallow Richards (1842–1911). Chem N Z 80:195–201.

© Springer Nature Switzerland AG 2020
B. Halton, *Some Forgotten Chemists*, Perspectives on the History of Chemistry, https://doi.org/10.1007/978-3-030-16403-4_14

Fig. 14.1 Ellen Swallow Richards—Vassar Class Picture—1870 (from Hunt CL (1912) The Life of Ellen Richards via Wikipedia)

enough money, but as there were no colleges open to women in New England she went to the recently opened Vassar College in Poughkeepsie on the Hudson River (some 125 km north of New York City). It had taken in its first class of 353 students paying $350 for tuition and residence just three years earlier, offering young women a liberal arts education equal to that of the best men's colleges of the day.

She started on September 17, 1868 with some $300 and as a 25-year old was a special student. Although co-educational, then it was a pioneer college for women's education in the eastern US that became one of the Seven Sisters colleges, the women's equivalent of the Ivy League. Vassar's main building was fitted for teaching sciences by its Professors Maria Mitchell (Astronomy and the first academic appointee) and Charles Farrar (Chemistry and Physics). These subjects and their teachers attracted Ellen and she devoted much time to studying astronomy under Mitchell, the most important woman scientist in 19th century America and one of the first women science professors. However, Farrar's influence led Ellen into chemistry with the idea that science should help in the solution of practical, everyday problems. Her immediate need was to meet the fees and because she had taught, Farrar allowed Ellen to tutor in the evenings.

Farrar seems to have given Ellen, one of his more ardent students, a good basic chemical education. Early in her second year he advised the students to take special care with their analytical work and be accurate as *'the profession of an analytical chemist is very profitable and means very nice and delicate work fitted for ladies' hands'* [1]. He was of the view that his students should be aware of all aspects of chemistry and planned a visit to the West Point Iron and Cannon Foundry for industrial experience but the college principal refused to let it happen as women on the workshop floor could give the institution bad publicity. Farrar provided a range of daily life examples such as 'strong' (rancid) butter containing butyric acid and not to be eaten, and the need for appropriate house ventilation (he burnt candles at different heights below an open window in an otherwise closed room, recording the time it took before they went out). Farrar encouraged Ellen through her time at Vassar, even addressing her as Professor Swallow a little ahead of her graduation with a BS on 21 June 1870.

Ellen Swallow took little notice of the advice to seek employment as an analytical chemist, but wanted more. For her immediate future she had secured a teaching post in Argentina under a government scheme, but the existing Paraguayan War worsened and the Argentinian government had to break the contract. As a result, she took what was probably her first holiday in years, returning to the area of her childhood, spending three weeks there and deciding to seek a career in chemistry. She wrote to the Boston commercial chemists, analysts and assayers, Merrick and Gray, seeking an apprenticeship. Merrick, a professor at the Massachusetts College of Pharmacy and consultant in the company replied to Ellen saying that they were unable to take her. Surprisingly for 1870, he suggested that she try to gain entry to the Institute of Technology in Boston (MIT) even though it did not take women. She wrote to the Institute knowing full well that gaining entry would be difficult but gave her Vassar professors as referees. As she received no early response she wrote to Booth and Garrett of Philadelphia as Booth's laboratory in Pennsylvania had taught chemistry for some 35 years. With associate Thomas H. Garrett joining in 1848, the firm had gained a strong reputation and, according to the *Scientific American*, a course in the laboratory was a necessity for the chemist of the time which it regarded of more value than a college diploma. The mid-November reply said that they had no vacancy, offered alternatives, and insisted that they '*desired to see proper means of livelihood thrown open to females*'. Swallow responded to this and it led to the offer of a studentship at a cost of $500 p.a. but urged her to gain entry to a scientific school if she could not afford it.

On Ellen Swallow's 28th birthday, her case, having been in the hands of the MIT secretary, was formally put to the faculty for its December 10 meeting. They recommended to the MIT Corporation (the Trustees)[1] the admission of Miss Swallow as a special student in chemistry but resolved that the admission of women as special students was an experiment with each case to be treated on its merits [1]. As it happened, women students were in a small minority at MIT (numbered in dozens) prior to the completion of the first women's dormitory in 1964. On December 14, 1870, the MIT President J. D. Runkle wrote to congratulate Ellen on her admission, asked her to come to see him and advised that '*you shall have any and all advantages which the Institute has to offer without charge of any kind*', this because he could then say she was not a student should anyone wish to object. It was not, as Swallow thought, a result of her financial plight.

In early January 1871, Ellen Swallow, already holding a BS degree entered MIT as a third-year student. She met with mathematician Dr. Runkle, President for 8 years from 1870 and strong supporter of her admission who was keen to see her succeed. The only woman working in the Roger's Building was the assistant in charge of

---

[1] On incorporation in 1861, MIT, as we know it, was the Massachusetts Institute of Technology and Boston Society of Natural History, but the original institute was known as the Boston Institute until it moved across the Charles River to Cambridge in 1916.

the chemical storeroom, Mrs. E. A. Stinson and with whom Ellen formed a strong long-lasting friendship. Stinson thought Ellen too frail and delicate to succeed in what was regarded as a difficult course and said so to Runckle. His response was: *But did you notice her eyes? They are steadfast and they are courageous. She will not fail* [1]. In settling to her life at MIT, Ellen Swallow changed little in her habits, determined to become a part of the institution. She continued her attention to routine chores but extended them to the college, e.g. she and Stinson swept a lecture theatre when the Janitor was taken ill, tidied staff areas and even sewed clothes and the like for the professors, all in addition to her classes. In essence she made herself an indispensable part of chemistry at MIT. She boarded at the house of a friend from Westford Academy throughout her student days, but catered and cleaned for herself to save money until her finances improved. Her ability as a student, her dedication to study and the accuracy of her laboratory work led to Professor Ordway (metallurgy and industrial chemistry, and a noted consultant on technical chemistry) letting her work for him, which previously had never happened with a student. It was during that first year that her father was struck by a train engine in the Worcester Union station and died shortly after. Ellen had been sick at the time and was at home. She remained there tending her mother for the remaining part of the 1871 teaching year. During the last months of the year she was supporting herself, settling her father's estate and making a daily return journey to MIT some 50 km distant to keep up with her study.

Returning to Boston in 1872 to continue her studies without the need for travel, she remained Ordway's assistant. However, in the summer of that year, Professor William R. Nichols, Head of Chemistry was engaged by the State to extend his earlier (1870) analysis of the waters of Mystic Pond and report on the quality from an industrial and sewerage pollution perspective. Although opposed to women's education, Nichols asked Ellen Swallow to be his assistant for this work accepting nothing short of absolute accuracy in her analytical results—the work was among the most advanced in sanitation (now environmental) chemistry in the world[2] [8]. She did this as well as the work for Ordway, her classes and study, and her general assistance to anyone. So impressed by her results, Nichols acknowledged in his report that it was she who had performed most of the analyses. This led to her becoming involved in the later (1887) State wide water survey (see below). Then Ordway's work on oils led to contact with Edward Atkinson, the inventor of the Aladin oven (see below). In 1873 Ellen H. Swallow graduated BS in chemistry with a thesis on '*Some sulpharsenites and sulphantimonites from Colorado*'. She also gained an MA from Vassar College that year having simultaneously worked on a thesis on the amount of vanadium in iron ore. That same year she acquired a 7 g black sample of the rare mineral samarskite that had reddish brown dark streaks and a vitreous lustre [9]. Samarskite is a radioactive rare earth mineral series that includes samarskite-Y: $(YFe^{3+}Fe^{2+}U,Th,Ca)_2(Nb,Ta)_2O_8$ or $(Y,Fe^{3+},U)(Nb,Ta)O_4$ and samarskite-Yb: $(YbFe^{3+})_2(Nb,Ta)_2O_8$. Her very careful

---

[2]The water analyses involved several species, e.g. Cl^-, Na^+, CO, H_2S and trace metals. The chloride analysis was much the same then as now, see Haywood and Smith [7].

analysis led to an insoluble residue that she said contained an element yet to be found and sometime later, samarium and gadolinium were isolated from the ore [10].

With her M.A., two BS degrees and experience in analysis, Ellen Swallow was suitably qualified for entry into a doctoral programme but women did not take such courses then and, despite her astonishing record, no one was willing to support the needed application; MIT did not graduate a woman with a Ph.D. until 1884. Undeterred by this she remained at MIT as a resident graduate continuing to conduct various analyses as earlier and devised a new method for determining the amount of nickel in various ores in 1877. She was essentially an unpaid chemistry lecturer at MIT from 1873 to 1878. During 1873 and the following year she was courted by the professor of mineralogy and assaying, and head of the department of mining engineering, Robert H. Richards, with whom she worked in the mineralogy laboratory. Richards proposed in the chemistry laboratory and they were married on June 4, 1875. They took their honeymoon in Nova Scotia—accompanied by the entire mining class that Robert was teaching! Robert Richards, who became professor of metallurgy in 1884, was a strong supporter of his wife's ambitions and she of his. They spent the summers of 1881 and 1882 in the copper regions of Northern Michigan where he was studying methods of concentrating and smelting copper and she his analyst. In 1879 she was recognised by the American Institute of Mining and Metallurgical Engineers as their first female member.

One of the first matters Ellen Richards pressed MIT for was a Woman's Laboratory. This followed from the Lowell Free Lectures and the increased need for laboratory instruction in science. Most teachers were women and there was a clear need for instruction for them. In fact, Ellen had taught a course for sixteen women at Boston Girl's High School with one of its female staff in 1873. Most were over 20 years of age and half were school teachers. To press her cause, Ellen appeared before the Woman's Education Association in late 1875 with the result that the MIT authorities were approached with an offer to equip a laboratory for women and provide the necessary books (the institution was in a dire financial state at that time). The result was for MIT to admit special students in chemistry without regard to sex in early May of 1876 and set up the Woman's Laboratory which opened in November under the charge of Prof Ordway with Mrs. Richards as assistant. It was housed in a portion of a one-story structure located between the sites of the Rogers and Walker Buildings. Over the following seven years, Ordway and Ellen worked in the lab without pay and Ellen, through her husband's support, donated an average of $US1000 each year to it. 1879 saw Ellen recognized by MIT, becoming an assistant instructor without pay, teaching the curriculum in chemical analysis, industrial chemistry, mineralogy, and applied biology. The laboratory was very successful. In 1883, MIT began awarding undergraduate degrees to women on a regular basis without need for special admission—and Mrs. Richards was heavily involved in ensuring that appropriate facilities for women were provided and in place. Moreover, a new laboratory had been created with space to accommodate all students irrespective of gender and the Woman's Laboratory was closed and removed when the Walker Building was erected. Her dedication to experimental science for women led her to becoming a founding member of the Naples Table Association for promoting laboratory research by women in 1898 (an equivalent to that set up in Naples, Italy with a small fully equipped laboratory [11]).

In 1884 MIT opened a separate laboratory for sanitary (environmental) chemistry with Mrs Richards as instructor, a position which she held until her death. Professor Nichols was head until his death in 1886 when Thomas M. Drown took over. For many years Ellen directed the entire instruction in the chemistry of air, water and foods for chemists, biologists and sanitary engineers, and only relinquished the chemistry of food supplies when the pressure of other affairs made this necessary. Then in 1886, the Massachusetts Legislature passed an Inland Waters Act to provide a comprehensive programme to protect the state's rivers, streams, and ponds. As part of the implementation of the Act, the Lawrence Experiment Station was founded in Lawrence in 1887, and headed by Professor Drown, whose principle mission was to develop practical methods for treating the growing volumes of wastewater that were seriously degrading surface waters in the state. The laboratory of sanitary chemistry was the world's first trial station for drinking water purification and sewage treatment. In 1887, the laboratory was asked to conduct a study of water quality state-wide for the Massachusetts State Board of Health. Ellen Richards, who had been chemist to the Board of Health from 1872 to 1875 was asked to carry out the analyses and became Board of Health water analyst from 1887 to 1897. The study took two years and involved monthly sampling from the chosen sites with more than 20,000 samples in total analysed as soon as possible after arrival at the lab. It was the first such study in America, and Ellen's data were used to find causes of pollution and improper sewage disposal. In carrying out their analysis of the data Richards and Drown generated what became known as the Normal Chlorine Map. All places with natural unpolluted waters that had the same chlorine concentration were joined by lines on a map and termed isochlors. When completed, it was found that the isochlors roughly paralleled the coastline and that the difference in chlorine concentrations between the different lines was similar to the distance of the line to the sea shore. The conclusion was that all places the same distance from the shore should have the same chlorine level and if above this it was due to pollution (the Cape Cod peninsula with water on all sides was excluded). The concept of the map and the way it was derived found much use worldwide. As a result of the study, Massachusetts established the first water-quality standards in America and its first modern sewage treatment plant at Lowell. The Lawrence Experiment Station grew into an engineering laboratory that performed pioneering research on the treatment of water supply, sewage, and industrial waste. Moreover, the practical engineering principles developed there led to dramatic reductions in water-borne diseases such as typhoid. Among many other works in sanitation was an investigation of the sanitary condition of the public-school buildings in Boston but little heed of it was taken.

Ellen Richards' dedication to her students was inspiring, and her personal and financial sacrifice for her pupils amazing.[3] The full effectiveness of her laboratory will probably never be adequately known as she also maintained an extensive private practice in sanitary chemistry for many years, and acted in an advisory capacity for

---

[3]Ellen Richard's contribution to her students is best described in Ref. [1].

a very large number of public and private institutions. Her publications relating to sanitation have been numerous and varied, and she maintained active membership in and participation at the meetings of local and national societies dealing with water supplies and public health problems. From her work in the new lab, Ellen Richards became one of the foremost sanitary chemists in the world.

Her time at the Women's Laboratory made Ellen Richards very meticulous in applying scientific principles to domestic situations. These included nutrition, pure foods, clothing, physical fitness, sanitation, and efficient home management that led to the creation of home economics—especially efficient practices that would allow women more time for pursuits other than cooking and cleaning. In essence it was a very early application of 'time-and-motion' with science to the home. When she had a gas oven installed in her home she also had a meter put in place so that she could record the amount of gas used to cook the meal, the recipe for it, and the length of time needed for preparation, thus giving an indication of the total cost of the meal. She also ensured that the house was appropriately ventilated and had a water back (wetback) installed for winter heating by the living room fire and a small heater in the basement for summer use. In 1882, she published *The Chemistry of Cooking and Cleaning: A Manual for Housekeepers* (see below).

Her interest in the environment led her in 1892 to introduce the word "oekology' (ecology) in the US which had been coined in German to describe the 'household of nature'. Together with Marion Talbot (an 1880 Boston University graduate) they became the founding mothers of the Association of Collegiate Alumni (ACA) in 1882, with the aim of opening the doors of higher education to other women and to finding wider opportunities for their training. She was elected an honorary life member in 1907. From the ACA came in 1883 the *Sanitary Science Club* comprised of some fifteen women graduates who met in the Richards' home to learn about home science, each studying their own home. It led to her 1898 booklet with Talbot on Home Sanitation. Furthermore, home study was promoted and Richards became one of its promoters, a forerunner of this century's correspondence school. The ACA was the forerunner of what is now the American Association of University Women (AAUW). It became a leading advocate for education and equity for all women and girls. Today, there are more than 100,000 members, 1300 branches, and 500 college and university partners nationwide.

In 1889, Ellen Richards set up the New England Kitchen which gained financial support from industrial manufacturers and other sources. It had an Atkinson Aladin oven installed and tested. The aim was to provide an open kitchen that prepared and cooked nutritious lunches that could be sold at cost, each with its contents and nutritional value provided. In many ways it was the first take-away outlet in the US. It opened on January 1, 1890 and led to other equivalent kitchens. By repeated chemical analysis, the methods of preparing a dish were perfected, giving a food whose consistency varied only slightly from day to day. One, with a constitution akin to milk fat was welcomed by the medical practitioners of the city.

The existence of the kitchen led some years later to hygiene requirements in restaurants. Success was daunted, however, by one lady gaining publicity for saying '*I don't want to eat what's good for me; I'd rather eat what I'd rather*'. Ellen cam-

	Food Value in Grams			
	Proteid	Fat	Carbo-hydrates	Calories
Voit's Standard. One-quarter of one day's ration	24.5	14.0	125.0	742.0
Atwater's Standard. One-quarter of one day's ration	31.2	31.2	114.0	882.0

	Ounces	Grams				
Baked Beans	8.4	238.1				
Brown Bread	4.2	119.1				
One Roll	2.0	56.7	26.3	35.6	131.4	979.3
Butter	0.7	19.8				
Apple Sauce	5.3	150.2				

Fig. 14.2 A Rumford Kitchen menu from 1900 (taken from Ref. [1], p. 221)

paigned tirelessly for the new discipline of home economics as it became known and pioneered an equivalent kitchen at the Chicago Exhibition of 1893—the Rumford Kitchen [12]. It was part of the Massachusetts State exhibition and laid bare all the cooking processes to the public. Visitors could buy any one of four different lunches each for 30 cents, the food value of which was carefully computed and provided on the bill of fare (a typical menu is shown in Fig. 14.2). The kitchen provided the first simple explanation and demonstration of the terms protein, carbohydrate, calorie, and that scientific methods underline nutrition. She became a consultant to a wide range of publicly funded state institutions that were required to cater to large numbers (hospitals, asylums, schools, etc.). An outcome of this endeavour was a successful plan to serve school lunches in Boston, another area in which Ellen Richards became the authority. She established programmes of study and organised conferences on home economics. Then, growing from discussions at several summer conferences that began in 1899 held at Lake Placid in New York State, the American Home Economics Association was formed in Washington DC in 1908 with Richards as its first president. In 1909 it established a Graduate School of Home Economics. W. O. Atwater, known for his studies of human nutrition and metabolism, said of Ellen Richards: *'The science of household economics is what chemists call a state of supersaturated solution; it needs only the insertion of a needle to start a crystallisation'*. Her efforts led to home economics becoming a subject of serious study for all. During this time, she introduced the word "euthenics" (eugenics by Francis Galton) to the language meaning the science of the environment controlled for right living, an area that has received increasing attention. Throughout her life and despite the numerous commitments she had, Ellen Richards continued to travel for her work and with her husband for his. Her last trips were to Mexico in 1901 and Alaska in 1903, each one with her portable water laboratory taking samples and outlining future study where needed. She was a model of efficiency.

At home, it was customary for the Richards to invite students who had little money to board in return for doing the housework. It was also customary for them to entertain their students giving dinner parties with occasional entertainment. Prof Richards taught glassblowing at MIT and often gave demonstrations after dinner. One of his more noted skills was to blow a water hammer much to the enjoyment of the guests. Later in life when in her 50s, Ellen eased her puritanical upbringing to the extent of developing an interest in theatre, which she and Richard took to attending on Friday evenings.

Richards served on the board of trustees of Vassar College for many years and was awarded an honorary doctor of science degree there in 1910. She served as a consultant to the Manufacturers Mutual Fire Insurance Co. from very early in her career. She died on March 13, 1911 at her home in Jamaica Plain, Massachusetts, now a National Historic Landmark (the Ellen Swallow Richards House). In her honour, MIT designated a room in the main building for the use of women students, and in 1973, on the occasion of the hundredth anniversary of her graduation, established the Ellen Swallow Richards professorship for distinguished female faculty members. Prof Robert Richards remarried in June 1912 (to Lillian Jameson) and remained married until her death in 1924. Surviving both wives, he lived to 101 years of age and died on March 27, 1945.

Ellen Swallow Richards was the foremost female industrial and environmental chemist in the United States in the 19th century, pioneering the field of home economics. She was a pragmatic feminist, as well as a founding ecofeminist who believed that women's work within the home was a vital aspect of the economy. There are few women, if any, alive today who would be able to devote themselves so totally and with as much energy as Ellen Richards displayed. The books, booklets and pamphlets she published, many of which are digitised for complimentary download are given separately following the citations.

A few of her writings and quotations appear below:

> Girls may learn that rice is a carbohydrate and peas and beans are not only carbohydrates but also albuminoids, without learning the connection of these facts with everyday life.

> Where are the fruits of chemical science? In the self-rising flour in bread powders, in washing powders, in glove cleaners, and in a hundred patent nostrums (medicines).

> If you keep your feathers well-oiled the water of criticism will run off as from a duck's back.

References

1. For a full biography see: Hunt CL (1912) The life of Ellen H. Richards. Whitcomb & Barrows, Boston
2. Sedgwick WT (1911) Ellen Henrietta Richards, A. M, Sc. D. Technol Rev 13:365–371
3. Swallow PC (2014) The remarkable life and career of Ellen Swallow Richards: pioneer in science and technology. Wiley, New York
4. Vassar Encyclopedia (2005) Ellen Swallow Richards. https://vcencyclopedia.vassar.edu/alumni/ellen-swallow-richards.html. Accessed 21 Jan 2019
5. MacLean M (2014) Ellen Swallow Richards. Civil war women. https://www.civilwarwomenblog.com/ellen-swallow-richards/ Accessed 21 Jan 2019
6. MIT archives. Index to material about Ellen Swallow Richards. https://libraries.mit.edu/archives/exhibits/esr/esr-bibliography.html. Accessed 21 Jan 2019

7. Haywood JK, Smith BH (1907) Mineral waters of the United States. Government Printing Office, Washington
8. Francis E (2010) Practical examples in quantitative analysis, forming a concise guide to the analysis of water, etc. 1873, Kessinger Legacy Reprints
9. Swallow EH (1875) Analysis of samarskite from a new locality. Proc Boston Soc Nat Hist 17:424–428
10. Adunka R, Orna MV (2018) Carl Auer von Welsbach: Chemist, Inventor, Entrepreneur. SpringerBriefs in History of Chemistry. Springer, Heidelberg
11. Sloan JB (1978) The founding of the naples table association for promoting scientific research by women, 1897. Signs 4(1):208–216
12. The Rumford Kitchen Exhibit at World's Columbian Exposition, Chicago, 1893. https://libraries.mit.edu/archives/exhibits/esr/esr-rumford.html. Accessed 22 Jan 2019

Books by E. H. Richards

The chemistry of cooking and cleaning; a manual for housekeepers. Estes & Lauriat, Boston, 1882
Food materials and their adulterations. Home Science Publishing Co., Boston, 1898
Home sanitation: a manual for housekeepers (with Talbot M). Home Science Publishing Co., Boston, 1898
The cost of living as modified by sanitary science. Wiley, New York, 1899
Plain words about food: the Rumford kitchen leaflets, 1899
Air, water and food from a sanitary standpoint (with Woodman AG). Wiley, New York, 1900
The cost of food: a study in dietaries. Wiley, New York, 1901
The dietary computer. Explanatory pamphlet; the pamphlet containing tables of food composition, lists of prices, weights, and measures, selected recipes for the slips, directions for using the same. Wiley, New York, 1902
First lessons in food and diet. Whitcomb & Barrows, Boston, 1904
The art of right living. Whitcomb & Barrows, Boston, 1904
The cost of shelter. Wiley, New York, 1905
Good luncheons for rural schools without a kitchen. Whitcomb & Barrows, Boston, 1906
Sanitation in daily life. Whitcomb & Barrows, Boston, 1907
Laboratory notes on water analysis. A survey course for engineers. Wiley, New York, 1908
The cost of cleanness. Wiley, New York, 1908
Euthenics, the science of controllable environment; a plea for better living conditions as a first step toward higher human efficiency. Whitcomb & Barrows, Boston, 1910
Conservation by sanitation; air and water supply; disposal of waste (including a laboratory guide for sanitary engineers). Wiley, New York, 1911

Chapter 15
Philip Wilfred Robertson (1884–1969)

Philip Wilfred Robertson (Fig. 15.1) was born in Ponsonby, Auckland, on September 22, 1884 the son of Donald Robertson, a postal clerk who was to become public service commissioner in Wellington, and his wife Edith Martin [1]. Called PW by most, though Robbie by his friends [2], was educated at Ponsonby School in Auckland and The Terrace School, on Fitzherbert Terrace in Wellington, New Zealand. He gained his secondary education at Wellington College where he was school dux in 1900 and then entered Victoria College to study the sciences. He took the introductory programme in chemistry taught by Professor T. H. Easterfield in the second year that it was offered. At that time the college had no permanent building of its own and science classes were held in the upstairs rooms of the Wellington Technical School on Victoria Street [3, 4]. After an outstanding undergraduate career, PW graduated MA with first-class honours in chemistry, conferred on the 29 June in 1905, and an M.Sc. in 1906 by which time he had published some nine papers alone in the journals of the Chemical Society (UK). Easterfield promulgated the concept of research from the earliest time of study, but even so, this level of publication by a student in his first years is truly amazing. PW also had papers read by Easterfield before members of the New Zealand Institute, the forerunner of the Royal Society of New Zealand, and published in their proceedings; one paper was published jointly with Easterfield.

As the leading student in his class PW gained a Sir George Grey Scholarship, a Senior Scholarship, and the Jacob Joseph Scholarship. His senior scholarship was the second for Victoria College. These achievements, coupled with his prowess at hockey and tennis, secured for him Victoria College's first Rhodes scholarship and the second offered in New Zealand. He proceeded to Trinity College, Oxford in October 1905 and won an open science scholarship there with a value of GBP 80. This scholarship was considered by the Oxfordians to require a greater scholastic feat than the winning of a Rhodes scholarship [5]. As PW was receiving the stipends of a Rhodes scholar, he was entitled only to the honour and glory of the open science

---

A previous version of this chapter was published as Halton B (2015) Philip Wilfred Robertson (1884–1969). Chem N Z 79:51–55.

Fig. 15.1 Left: P. W. Robertson, 1950 (unattributed from *J. NZIC*, 1950, *14*, 68) right Robertson lecturing in the 1940s (from a photograph by the late Des Hurley) (both with permission)

scholarship apart from certain privileges attached to the title of Scholar of Trinity. Nonetheless, the college authorities chose to give him the benefit of the scholarship by allowing him to pursue his studies abroad for a year at the close of his Oxford course.

Robertson gained first-class honours in natural sciences after two years and spent a year pursuing other chemical interests that led to further publications. Then, in May 1908, he registered at the University of Leipzig to take his Ph.D. with Professor Arthur R. Hanszch. Hantzsch is the well-known heterocyclic chemist whose pyridine and pyrrole syntheses were each named after him. PW was a competent linguist having studied German under von Zedlitz at Victoria College and, it is reputed, was so fluent that he easily passed for a native in Germany and Austria. His research in Leipzig involved the salts and hydrates of hydroxy azo compounds and copper ammonia complexes. He gained three publications with his supervisor in Berichte der Deutschen Chemischen Gesellschaft, one in 1909 and two in 1910. His PhD was completed in June 1909 with the title (in German) *Optical Studies of I Copper ammonia complexes, II Yellow and red salts of oxy azobenzenes*. In those days, formal application for doctoral examination was required and PW applied for his on June 9, 1909, as shown. It was followed by a letter asking if the likely date of examination was known. It actually took place on June 22 when PW was deemed to be an outstanding first-class candidate.[1] Among the courses that he took during his Ph.D. were those on the photochemistry of organic compounds, mineralogy, salts and complexes, and theoretical and technical electrochemistry, all from the Chemistry Department staff, Böttger, Le Blanc, Ley, Stobbe and Zirkel.

Following his doctoral completion in Leipzig, Robertson accepted appointment as professor of chemistry at Government College, Rangoon, Burma. Prior to 1904 this was Rangoon College but that year its name changed to Government College and stayed as such until 1920 when it became University College ahead of merging

---

[1] I thank Petra Hesse of the University of Leipzig archive for providing this information. Unfortunately the thesis is no longer available.

15 Philip Wilfred Robertson (1884–1969)

that same year with the Baptist-affiliated Judson College. By 1880, Arts and Science intermediate level courses at Rangoon College were given by staff members of the College and recognized by London University. Robertson became one of these. His decision to take a position in Burma was to enable him to study Buddhism first-hand as he has described in his autobiography [6]. The sojourn in the tropics was less than two full years, but during his time there he contracted malaria and, although he recovered, its effects never completely left him to the extent that he was often cold, muffled in a thick scarf and overcoat even on the warmest of days, and he rarely went out in the evenings [7].

From Rangoon, PW returned to England in 1911 to a lectureship at Imperial College London. His time there generated six further publications with the research students that he supervised. It was after his return to NZ in 1912 that PW married Florence Elizabeth Graham, whom he met in London. Florence was a widow with a seven year old child Sybil, and it was only in 1917 that the couple had their own child, daughter Monica born on October 5 that year.[2]

When Thomas Easterfield, Victoria College's inaugural Professor of Chemistry, resigned to take up the Directorship of the newly established Cawthron Institute in Nelson, the selection of his successor seems to have been easy. Robertson had been an outstanding early student who had established himself in London and had just been awarded the 1919 Hector Medal by the New Zealand Institute; he was the second chemist (after Easterfield) to receive this award. It is not surprising, therefore, that Robertson was offered the chair in chemistry and his acceptance of it was announced by the College Council on December 17, 1919 and reported in The Evening Post the day after [9]. As an aside, he became the fourth member of the College professorial board to hold a New Zealand university degree. The return to his *alma mater* was for the 1920 academic year and he arrived with his wife, stepdaughter and daughter. The Robertson household was established in Talavera Terrace (below Weir House) and next door to Professor T. A. Hunter, Victoria's Professor of Philosophy. PW remained as Professor of Chemistry at Victoria for some 30 years.

Robertson's arrival in Wellington coincided with the expansion of the College's sole building, now known as the Hunter Building. Government had approved an expansion to the north to accommodate the library, but it was about the time of PW's arrival that the announcement for a science expansion of the south of the building was made [4]. Thus, his first three years at Victoria University College, as it then was, was dedicated to the building expansions with the noise, dust and dirt that go with them. On arrival, the only employee in chemistry was George Bagley, the Assistant and Demonstrator and a former Easterfield student. During his first year PW had A. D. (Bobby) Munro appointed to follow Bagley, a position that subsequently became permanent and to which Munro was appointed in 1928 as the second academic in chemistry—a lecturer. PW continued the previously set teaching programmes that included evening and Saturday lectures to which he added a tutorial after the Friday evening discourse. His impact on the College calendar was evident in 1922 as the

---

[2] I am grateful to John Milner, Monica's son and P. W. Robertson's grandson for this and other matters of family history. See also Nicholls [8].

course, times, and offerings became more regulated. By then research was performed for an M.Sc. in the sciences and it was from the work of students from this pool that most of PW's subsequent New Zealand research was advanced. No Ph.D. degree in chemistry was awarded in Wellington until the late 1940s despite a short trial period in the mid-late 1920s, but that saw no chemistry candidate.

The extensions to the Hunter Building were completed in 1923 and formally opened on October 23. By then PW's textbook *Quantitative Analysis in Theory and Practise*, written in London with D. H. Burleigh, had been published and become a standard text in many universities [10]. As a lecturer, PW's impact on the student body has been recorded by a few of those who attended his classes from the late 1920s until the mid-1940s.[3] PW was reported as entering the lecture theatre though a door beside the bench at which time the buzz in the room stopped abruptly. He was tall and sparely built, always wearing his small, steel-rimmed glasses and, up to the time of his retirement, a brown suit with high, white, Edwardian stiff shirt collar and a paisley tie. He would approach the lectern and place his papers on it without looking at the students until he had seated himself on a tall laboratory stool behind the bench, folded his arms and cleared his throat. Only then did he glance briefly at the class before launched into his discourse. The first lecture of the year began with *Chemistry consists of two main branches* [3]. He spoke, matter-of-factly, though quietly, pronouncing his words clearly. Occasionally he would rise to turn and write a new chemical term on the blackboard before returning to his perch, erect on the stool. Occasionally, when he was unwell from his malaria, his assistant would place a small table and chair for him and on cold days he would light the Bunsen burner, which was always on the bench, but never used for demonstration. The material of each lecture was painstakingly condensed, and frequently updated from the latest developments provided in the literature.

Underneath his controlled exterior PW had a whimsical sense of humour, very dry and not often apparent, but it could show itself unexpectedly as a funny remark in the middle of a lecture. Because he rarely changed his delivery style, the class frequently missed the point until several sentences later when someone would laugh belatedly, to be followed by the rest of the class. A thin smile would then pass across PW's face. When demonstrating in the laboratory, PW moved along the rows of students methodically, guiding each one in turn. He would perch on the student's stool, watch their manipulations and offer gentle correction when it was needed, often by way of outrageous analogy never to be forgotten by the student. For example, on one occasion a student used too much water to wash a precipitate free of contaminant and much of the needed precipitate was lost. PW had sat behind watching with folded arms. He said nothing until the procedure was finished and the female student was glumly observing the small quantity of material remaining. Without emotion, PW quietly remarked: *You know, it is easier to wash a baby with several thimblefuls of water than a bath-full* [3].

---

[3] I am indebted to the late Brian Shorland (see Ref. [3]), Joan Cameron and Bruce Cockburn for their comments.

15 Philip Wilfred Robertson (1884–1969)

Philip Robertson's wife, Florence, did not settle well to life in Wellington to the extent that in 1927 she took her two daughters and returned to England and had Monica educated at boarding school from 1930 to 1934. Apparently, Florence returned to Wellington in 1930, but never for very long periods until after Monica completed secondary schooling. Her initial departure left PW an involuntary bachelor and he shared his home with philosophy lecturer (and subsequently noted psychologist) Ivan Sutherland [11]. However, with her 1930 return, Ivan was forced to move out [11]. After Florence and Monica returned to Wellington in 1934 Monica attended Victoria University College for a couple of years, presumably from the commencement of the 1935 academic year. However, PW dissuaded (even prevented her) from completing her degree as he felt it inappropriate for women to be educated to that level! Monica and Florence returned to the UK permanently in 1937 or 1938 by which time Robertson's marriage had ended. Monica married Bill Williams, subsequently editor of the *Dictionary of National Biography*, whom she had met in Wellington in 1938, but that did not last as he returned from a distinguished army career in WWII a different man. Monica subsequently recorded that Florence apparently did not care much for her husband [7] Florence died in 1968, one year before PW.

Philip Robertson firmly established the ethos of research in the Chemistry Department at Victoria University College becoming one of the few New Zealand chemists of his generation to establish a lasting international reputation for his scientific work. It was achieved with minimal equipment and funding but with the research assistance of 102 M.Sc. students. His work has been loosely described as encompassing the relationship between chemical constitution and chemical properties [12] and it was quite distinct from his doctoral studies in Leipzig. His undergraduate researches and his studies in London provide the foundations of much of his later work. The late Brian Shorland [3], a student of Robertson's from 1928, wrote of the works under five categories in a tribute to the man upon his retirement [12] and Peter de la Mare commented in his 1969 obituary in *Chemistry in Britain* [13]. In today's age the studies are best described as physical organic in nature, a classification that had not emerged in the era when Ingold and Hughes were advancing the concepts in the UK and Robertson doing the same in New Zealand, specifying it as kinetic in nature. It needs to be remembered that in the Robertson era supplies of chemicals and the equipment to carry out research were strictly limited to the extent that the favourite piece of equipment he had was a one-half ounce bottle capped with a ground glass stopper [12].

Thus, Robertson's studies can be described as involving analytical aspects which led to his elegant method of determining chlorine, bromine and carbon in organic compounds, the halogenation of phenols and the production of substituted aromatics, the physical properties of solutions (especially phenols), the melting characteristics of acid amides, and detailed halogenation studies [12, 13]. The late Joan Cameron described his achievements in chemistry as '*including substantial contributions to the basic understanding of the reactivity of certain carbon compounds, and how their properties such as boiling points were related to their chemical structure*' [3]. Studies by Robertson and his students extended the knowledge of melting point variations in homologous series of compounds to comparable variation in compound association thereby laying the foundation for structural differences and compound configuration. These conclusions were enhanced by observations in the amide series covering a wide range of compounds. The attention to detail that Robertson instilled in his charges was such that his tabulation of melting points appeared in an early text on the fatty acids and their derivatives [14]. It was the relationship of physical properties and compound composition that led PW to his major interest in organic reaction mechanisms.

The halogenation work of Robertson attracted most attention and led to some 28 publications. He added substance to Ingold's theory of electrophilic halogenation and provided rate data to substantiate the acceleration cause by the presence of electron donating alkyl substituents and the reduction by electron accepting groups. His students showed that as electron withdrawal increased the reaction became nucleophilic in nature and the rate then increased. They also demonstrated that the reaction profiles are not always simple with solvent effects and pH giving way to ter- and tetramolecular processes. Their work expounded the mechanism of the reactions and in the latter case the presence of Br_4 was proposed [15] and HBr_3 in the nucleophilic reactions with HBr [16, 17]. The area was nicely placed in perspective by the late Peter de la Mare during his tenure as assistant lecturer at University College in London [18]. It was de la Mare's 1941 M.Sc. studies with Robertson that continued through fruitful collaboration to mid-1953, which not simply held Robertson's attention, but set de la Mare on the road to his distinguished career as a physical organic chemist and foremost authority on electrophilic substitution. Yet at the conclusion of his Victoria studies, de la Mare joined the Agricultural Research Laboratory to work with Brian Shorland. However, the enforced move of most of the laboratory staff to the Ruakura Research Station in Hamilton immediately post-WWII was not to his liking. Dissatisfaction with the new regime led Peter (among others [3]) to move to London for successful Ph.D. studies with (Sir) Christopher Ingold and a remarkably successful future career.

Philip Robertson played a major role in developing science in New Zealand by training the number of chemists that he did. A remarkably high number of his M.Sc. students went on to distinguished careers in science, many gaining a doctorate in overseas laboratories (dominantly the UK). Among those noted by Robertson himself in his 1949 review of science at Victoria College [19] were (M.Sc. graduation year in parenthesis): Dr. H. L. Richardson, adviser in soil conservation to the Chinese government (1925); Dr. G. M. Richardson, noted for his chemistry of bacterial processes at the Dunedin Medical School (1925); Dr. F. B. Shorland, head of the fats research division of DSIR who established the industry of extracting vitamin-rich fish oils and became a renowned nutritionist (1932) [3]; and Dr. P. B. D. de la Mare (1942), about whom Robertson in 1949 said '*at present lecturer at the University College, London, for whom a distinguished career in chemistry is confidently predicted*'. Peter de la Mare had gained a lectureship at University College London prior to the becoming Professor of Chemistry at Bedford College and before accepting the headship of chemistry at Auckland University in 1967. Others that now fully justify mention are O. H. Keys, inaugural editor of the NZIC journal (1931); T. A. Glendenning, tutor Wellington Technical College and second editor of the NZIC journal

Fig. 15.2 Christopher Perkins, *Portrait of Professor P. W. Robertson*, 1930, oil on board, with the permission of the Victoria University of Wellington Art Collection

(1921); N. T. Clare, chief scientist Ruakura Animal Research Station who had been PW's lab boy (1934); B. E. Swedlund, a reader in chemistry at Auckland University (1944); E. P. White, the agricultural chemist who isolated, purified and structurally identified spirodesmyn (1938); I. K. Walker, a Director of the DSIR (1937); H. P. Rauthbaum, distinguished DSIR scientist (1947); W. E Dasent, lecturer, bursar and registrar at Victoria University (1950); and J. K. Hayes, Professor of Botany at Victoria University (1952). In addition to his students' achievements, PW was involved in the early planning of Victoria's new building to the south of the Biology Block projected for Chemistry and Geology—subsequently to be named the Easterfield Building, which gained priority in the developmental schemes of the College and became the first science building in the post-WWII era in New Zealand.

Unlike the vast majority of his scientific colleagues in New Zealand, Professor Robertson was a cultured individual who gained much recognition in literary circles [2]. He became friendly with the British artist Christopher Perkins and socialised with him and his family in their home at 151 Upland Road in Kelburn, and in his own. It was on his second meeting with Perkins that PW commissioned the artist to paint his portrait (Fig. 15.2) [7], a painting that now hangs close to the School of Chemical and Physical Sciences in Victoria University of Wellington. PW paid Perkins the sum of GBP 20 for the commission and subsequently had him also paint a portrait of Professor George William von Zedlitz, Victoria's first professor of modern languages. PW had used his time in Oxford to become widely read, well travelled, and with an eloquent writing style. He believed that science should inform art, and art science as both were aspects of the whole. Thus, he was to the fore with his literary writings, which included his early *A soul's progress: mezzotints in prose* (1920) that represented five periods in the history of the soul of an imaginary young scientist trying to escape a narrow intellectual view of the world. He wrote letters to the local newspapers complaining about the poor status of the arts, criticised the government's lack of skill in managing the economy and its failure to support for the country's cultural heritage. He joined with other members of the professorial staff in public debates, in particular on the role of the university, freedom of thought and of the press.

P. W. Robertson was appointed emeritus professor following his retirement in late 1949 and returned to London shortly afterwards where he was closer to his daughter. He became a regular visitor to University College London. He died in London on May 7, 1969, aged 84.

References

1. Davis BR (1998) Robertson, Philip Wilfred. Dictionary of New Zealand Biography, Te Ara—the Encyclopedia of New Zealand. www.TeAra.govt.nz/en/biographies/4r23/robertson-philip-wilfred. Accessed 18 July 2014
2. Anon (1950) P. W. Robertson MA, MSc (NZ), MA (Oxon.), PhD (Leipzig), FRSNZ. J N Z Inst Chem 14:68–71
3. Cameron J (2014) Brian Shorland: Doyen of New Zealand Science. Curtis NF, Halton B (eds). New Zealand Association of Scientists, Wellington
4. Halton B (2018) Chemistry at Victoria—The Wellington University, 3rd ed. Victoria University of Wellington, Wellington. Available for free download at: https://www.victoria.ac.nz/scps/about/attachments/chemistry-at-victoria-third-edition.pdf
5. A New Zealander at Oxford. The New Zealand Herald, Volume XLIII, Issue 13088, 30 January 1906, p 5. https://paperspast.natlib.govt.nz/newspapers/NZH19060130.2.34. Accessed 22 Jan 2019
6. Robertson PW (1931) Life and Beauty—a spiritual autobiography. Edward Arnold, London
7. Garrett J (1986) An artists daughter with Christopher Perkins in New Zealand. Shoal Bay Press, Auckland
8. Nicholls CS (2004) Williams, Sir Edgar Trevor [Bill] (1912–1995). In: Oxford Dictionary of National Biography. Oxford University Press, Oxford
9. Chair of Chemistry. The Evening Post, Volume XCVIII, Issue 146, 18 December 1919, p 5. https://paperspast.natlib.govt.nz/newspapers/EP19191218.2.53. Accessed 22 Jan 2019
10. Robertson PW, Burleigh DH (1920) Quantitative analysis in theory and practise. E. Arnold, London
11. Sutherland O (2013) Paikea: The life of I. L. G. Sutherland. Canterbury University Press, Christ-church, New Zealand, p 195
12. Shorland FB (1950) The contributions of Professor P.W. Robertson to chemical research. J N Z Inst Chem 14:71–78
13. de la Mare PBD (1969) Philip Wilfred Robertson. Chem Brit 5:525–526
14. Ralston AW (1948) Fatty acids and their derivatives. Wiley, New York
15. de la Mare PBD, Robertson PW (1948) The kinetics of aromatic halogen substitution. Part IV. The 1-halogenonaphthalenes and related compounds. J Chem Soc, 100–106
16. Morton ID, Robertson PW (1945) The kinetics of halogen addition to unsaturated compounds. Part V. The $\alpha\beta$-unsaturated acids and hydrogen bromide catalysis. J Chem Soc, 129–131
17. Swedlund BE, Robertson PW (1945) The kinetics of halogen addition to unsaturated compounds. Part VI. The allyl halides: lithium chloride and hydrogen bromide catalysis. J Chem Soc, 131–133
18. de la Mare PBD (1949) Kinetics of thermal addition of halogens to olefinic compounds. Q Rev Chem Soc 3:126–145
19. Robertson PW (1949) Science at Victoria College. The Spike: or, Victoria College Review May: 26–28

Chapter 16
Women Pioneers

Historically, chemistry is a woman's science—more than any other is. The first known chemists were women whose names appear on the cuneiform tablets of ancient Mesopotamia some 4000 years ago. They were concocters and purveyors of perfume, the vital commodity in the age before personal hygiene. From then until the time of the Enlightenment in the 16th century, the secrets of medicines, cosmetics and perfumes passed from mother to daughter through the generations. Although the act of formally writing down and debating scientific knowledge only took hold of mainstream aristocracy in the 17th century, Mary the Jewess (Mary Phrophetissima aka Mary the Prophetess) was an alchemist who lived between the first and third centuries AD. She was one of the first writers and first true alchemist of the Western world [1]. The subject of chemistry and all sciences became male dominated from the Enlightenment. Nonetheless, Mary is credited with the invention of apparatus, that includes the bain-marie (Mary's bath) and the first true distillation still that consisted of copper tubing, ceramic pottery and metal.

According to Wellington et al. [2], the first Ph.D. was awarded in Paris in approx. 1150 but it was not until the early nineteenth century, following university practice in Germany, that the term Ph.D. acquire its modern meaning as the highest academic doctoral degree [2, 3]. The earliest record of a woman gaining an advanced degree came in the early 17th century with Spaniard Juliana Morell the first female to receive the degree for study on Aristotle [4–6]. She received her doctorate in Canonic Law in Avignon, France, in 1608. Her public defence was in the papal palace of the vice-legate before a distinguished audience. Later that year she entered the convent of Sainte-Praxède at Avignon. Then in 1678, Elena Cornaro [7], an Italian philosopher of noble descent gained her Ph.D. degree from the University of Padua and this is often cited as the first to a woman. Because of her gender Gregorio Cardinal Barbarigo, the Bishop of Padua, had refused to allow her to submit for a theology degree. However, he did permit a degree in philosophy and, after brilliant study, she received the doctoral degree and took up a lectureship in mathematics at the University that same year. Elena was an expert musician in addition to mastering almost the entire body of knowledge. She was proficient with the harp, harpsichord, clavichord, and the violin, and demonstrated her skills from the music she composed in her lifetime. She was a member of various academies and esteemed throughout Europe for her attainments and virtues.

The earliest acknowledgement of a women chemist that this author has found came from the early 17th century with Martine Bertereau, Baroness de Beausolei (1590–1643), a French mining engineer and mineralogist. Together with her husband, Jean de Chastelet, they travelled extensively in Hungary, Germany and South America in search of mineral deposits using divining rods and the like [8]. In 1626, they were commissioned by King Henry IV to survey France for possible mine locations and revive the French mining industry. They surveyed hundreds, subsequently in the service of Henry IV's son, King Louis XIII. However, religious and financial concerns led the government to move against them, bringing charges of witchcraft. Both husband and wife (together with their daughter) were separately imprisoned (the Bastille and Vincennes) where they died.

In 1656, the first chemistry book by a woman, *La Chymie Charitable et Facile, en Faveur des Dames* (Useful and Easy Chemistry, for the Benefit of Ladies) was authored by Marie Meurdrac. She is likely the most important female chemist and alchemist never heard of. Her tome discussed instrumentation (vessels, lutes, furnaces, weights), how to make medicine from plants (especially by purifying through distillation), as well as animals, metallurgy and compound chemistry [9]. The final component addressed the female audience and covered methods of preserving and increasing beauty. Irrespective of whether her work is chemistry, alchemy or medical cookery, Meurdrac directly contributed in print a means that allowed for collaborations, as well as scrutiny, and it later defined the field of modern chemistry and science as a whole. This book appeared in 1656, five years before Boyle's *The Sceptical Chymist*.

The first woman to gain a Ph.D. degree in science was Laura Maria Caterina Bassi (1711–1778) [10], who holds a unique place in science history and education. She gained her Ph.D. in physics in 1732 from the University of Bologna and that same year became the first woman to accept an official, salaried teaching position there. This was when, generally, women could not pursue studies and intellectual professions. Throughout her academic career, Bassi led a relentless struggle to achieve equal conditions in teaching, while making a name for herself in the world of academia, a world that then in Italy and worldwide was exclusively male. She also had a pivotal role in spreading Newtonian physics in Italy and in pioneering research on electricity. In Bologna, she presented dissertations on subjects such as gravity, refrangibility, mechanics, and hydraulics. With her husband, Giuseppe Veratti, she made Bologna a centre for experimental research in electricity. Despite numerous public appearances, most of Bassi's teaching and research took place in her home, where from 1749 until her death, she and her husband established a laboratory, taught classes in experimental physics and natural philosophy, and presided over a lively scientific "salon." In 1776, she accepted the chair of experimental physics at the Bologna Academy of Sciences, with her husband as her assistant and intellectual partner. He took it over in 1788 when she died.

Bassi's graduation was followed in Germany by the award of Ph.D. degrees in medicine to women and by a second physics Ph.D. from Bologna to Italian Cristina Roccati in 1751; she then taught physics at the Scientific Institute of Rovigo for 27 years. The woman chemist received no further recognition until after these events and into the last quarter of the 18th century. This came from the work of Marie-Anne Pierrette Paulze. Married at age 13 to Antoine Lavoisier, Marie-Anne began to assist her husband and, as her interest evolved, two of her husband's assistants, Bucquet and Gingembre tutored her. The Lavoisiers spent most of their time together in the laboratory, working as a team on many aspects of research, she as his assistant and translating chemical papers from English to French for her husband. Lavoisier, regarded as the father of chemistry, published the first accepted chemistry textbook in 1789, *Traité Élémentaire de Chimie* (*Treatise of Elementary Chemistry*) this some 120 years after the book by Meurdrac. After Lavoisier's death in 1794, Marie-Anne subsequently married Benjamin Thompson, Count Rumford, one of the most noted physicists of the era.

In 1794, some six months before Lavoisier's death, Mrs Elizabeth Fulhame published a book: *An essay on combustion: with a view to a new art of dying and painting*. This woman is another whose chemistry did not gain the recognition it deserved [11]. Married to Thomas Fulhame, an Irish-born physician who had attended the University of Edinburgh and studied chemistry, little is known about her other than that she appears to have been Scottish. She began her chemistry with an interest in finding way of staining cloth with heavy metals under the influence of light. She was encouraged to publish an account of her fourteen years of research after a meeting with Sir Joseph Priestley in 1793. She had studied the reduction of metallic salts in a variety of states (aqueous, dry, in ether and in alcohol) by exposing them to the action of various reducing agents, employing hydrogen, phosphorus, potassium sulfide, hydrogen sulfide, phosphine, charcoal, gas, and light. She discovered a number of reactions in which salts reduced to pure metals. Likely, her most important discovery was that of aqueous chemical reduction at room temperature, rather than from smelting at high temperatures.

Her theoretical work on catalysis was a major step in the history of chemistry, predating both Berzelius and Buchner, but for which she was given little recognition. She proposed, and demonstrated, that many oxidation reactions occur only in the presence of water, involve water, and that it is regenerated and detectable at the end of the reaction. Furthermore, she proposed modern mechanisms for the reactions, and may have been the first scientist to do so. The role of oxygen, as she describes it, differed significantly from other theories of the time. Thus, she disagreed with some of the conclusions of Lavoisier and the phlogiston theories that he analysed. Her research was a precursor to the work of Berzelius, although she focused on water rather than heavy metals. Fulhame's work on silver chemistry is a landmark in the birth and early history of photography and that on the role of light sensitive silver salts on fabric (photoreduction) predates Wedgwood's more famous photogram trials of 1801. Mrs. Fulhame was appointed a corresponding member of the Chemical Society of Philadelphia in 1810 which stated: *Mrs. Fulhame has now laid such bold claims to chemistry that we can no longer deny the sex the privilege of participating in this science also.*

Some 12 years after the Fulhame book, likely the first chemistry textbook appeared, the amazingly popular *Conversations on Chemistry* was published appearing anonymously in 1806. Its authorship was not revealed as Jane Marcet (1769–1858) until the 12th edition in 1832 [12]. Her book (in two volumes) served as a model of successful "popular science" writing despite its subject matter and was one of the first elementary science texts. It appeared in seventeen British and two dozen American editions, and numerous translations; 20,000 copies had sold in England by 1865. Humphrey Davy helped popularize it in his Royal Institution lectures and Michael Faraday gained his grounding in chemistry from it. Jane Marcet, the daughter of a London banker, had married Alexander John Gaspard Marcet, a doctor and subsequent lecturer at Guy's Hospital in 1799. He set up a home laboratory where his wife gained her chemical knowledge [13]. The book takes the form of conversations between two young female pupils, Caroline and Emily, and their teacher Mrs Bryant. The younger Caroline asks flippant questions that move the dialogue along, while the more controlled Emily is somewhat reflective. The maternal Mrs Bryant is a mentoring figure who leads them to question and examine their ideas [14].

Just two years later in 1808, Anna Sundström (1785–1871), the daughter of a farmer and born Anna Christina Persdotter who took the name later, became housekeeper and assistant to the famed Jöns Jacob Berzelius. From simply a housekeeper, she began to assist Berzelius and over a period of 28 years gained her chemical knowledge from working alongside him. She became an effective assistant and co-worker acquiring a vast knowledge of chemistry. Berzelius stated *"She is used to all my equipment and their names to such a degree that I could without hesitation make her distil hydrochloric acid"*. She maintained the laboratory and supervised his students, and was termed affectionately *strict Anna* by them. Her employment ended in 1836 after 56-year old Berzelius married Elisabeth Poppius some 32 years his junior. Sundström is immortalised in an annual award of the Swedish Chemical Society's inorganic chemistry division for the best Swedish Ph.D.-thesis—the Anna Sundström Award.

The first Ph.D. degree in chemistry awarded to a woman was still half a century away. Between 1837 and 1857 the education of women advanced significantly with co-educational and single sex schools opening throughout much of the northern hemisphere, Egypt and Chile [3] Bedford College opened in London in 1849 as the first higher education college for women in the United Kingdom while, in 1855, the University of Iowa became the first US university to become co-educational. In the late 1860s Anna Fedorovna Volkova gained chemical knowledge from lectures at St. Petersburg University and became the first woman to publish research in a chemistry journal (from work at the St. Petersburg Technological Institute) having two papers on sulfonic acids published in 1870 [1]; and working dominantly with amides. She became the first woman member of the Russian Chemical Society. Volkova was also the first to prepare p-tricresol phosphate from p-cresol, the o-analogue of which is an important plasticiser.

The universities in Switzerland were the first in modern-era Europe to admit female students. Lydia Sesemann (Lidiia Zesemann) from Vyborg in Finland was the first woman to study chemistry in the Faculty of Natural Sciences in Zurich (1869–1874) [15, 16]. She defended her dissertation in organic chemistry *on dibenzylacetic acid and a new synthesis of homotoluic acid* and gained her degree in May 1874 (see Fig. 16.1) for work under Victor Merz and Wilhelm Weith. She was the first woman to graduate in science at the university, but she never practised the profession, which likely accounts for her absence in much documented history. Many of the references to the early Ph.D. graduates accord Stefania Wolicka as the first woman Ph.D. graduate, but this is not so. Her university of Zurich matriculation document No. 3798 (the equivalent of Fig. 16.1) gives the date of submission (November 5, 1874) and award (March 6, 1875) [17], which are clearly after those for Sesemann. The first woman to study and graduate Ph.D. from Zurich was Russian Nadzhda Suslova who gained her degree in medicine in 1867 and was the first in Europe in the modern era [18]. The Russian Iuliia (Julia) Vsevolodovna Lermontova was the second Ph.D. chemist. The University of Göttingen granted her degree in the autumn of 1874 from studies with Bunsen in Heidelberg and Hofmann in Berlin at about the same time as her friend and compatriot Sofia Kovalevskaia (1850–1891) obtained her doctorate, the first woman Ph.D. in mathematics. Göttingen accepted these women for advanced degrees whereas neither Heidelberg nor Berlin would. Lermontova subsequently worked in Markovnikov's Moscow laboratory on aliphatic hydrocarbons and published alone (as J. Lermontoff) the synthesis of 1,3-dibromopropane (1876) [19] and 2,4,4-trimethyl-pent-2-ene (1879) [20], prior to working with Butlerov [21] in St. Petersburg. On returning to Moscow, she became likely the first woman to work on petroleum chemistry (Caucasian) with Markovnikov.

A year after Sesemann graduated in Switzerland, Lovisa (Louise) Katarina Hammarström (1849–1917) (Fig. 16.2) became the first formally trained female chemist in Sweden. Her father, a medically trained priest, was widowed when Louise young and she grew up at an Ironworks in Dalarna (central Sweden). There she became interested in chemistry from the city chemists she met. She was a student at Konstfack (now the University of Arts, Crafts and Design in Stockholm), then the Handicraft School, and studied chemistry by private lessons, though she never gained a formal qualification. From 1875 to 1881 she was employed an assistant in the engineering laboratory of Werner Cronquist in Stockholm, performing arsenic analyses. She then joined the Ironworks at Bångbro (1881–87) as a mineral chemist, then the Fagersta Bruk (1887–91) that produced carbon steel wire and rod, and finally at Schisshyttan (1891–93). In 1893, she opened her own Bergchemical laboratory in Kopparberg and focused on minerals and geological studies.

Nadezhda Olimpievria Ziber-Shumova (1856–1914) was another of the Russian women to be educated in Switzerland in the early 1870s. However, she was there together with her husband Nokoli Ziber (an early Marxist economist) and never enrolled for a degree in Zurich. When the Russian women were ordered out of the city by the Russian government in 1873, she transferred to Bern. By 1877, she was working with Polish chemist/biochemist, Professor Marceli Nencki, but she was still without qualification. When Nencki transferred to Moscow in 1891 to head the newly

Matrikeledition

Sesemann (Frl.) Lydia (Maria)

Matrikelnummer	3617
Fakultät	philos.
Semester	Wintersemester
Immatrikulationsjahr	1869
Geburtsdatum	
Geschlecht	weiblich
Herkunftsort bzw. CH-Bürgerort	Wiborg
Herkunftsort bzw. CH-Kanton	Finnland
Herkunftsland	russ.Reich
Angaben zur Vorbildung	unklar ob Stuttgart (= ev.Wohnort d.Eltern) od.Priv.unterr.in Finnland ?
Weggang von der Universität	ab mit Zgn.24.04.1874, (prom.15.05.1874)
Informationen über die Eltern	F., majorenn
Zusatzinformationen	* 1846, Dr.phil.II (StAZ U 110 c.1 = 1.Promotion einer Frau in Zürich an die- s Diss."Üb.Dibenzylessigsäure und eine neue Synthese der Homo- toluylsäure" (' S.34; MB

© Universität Zürich | 01.12.14 13:35 | Impressum

Fig. 16.1 Inset: Lydia Sesemann first woman doctor from Zurich University, Switzerland and her 1846 Zuruch matrekeledition (courtesy of Dr Irène Studer-Rohr, Chemistry Department, University of Zurich)

established, and first, Russian biochemistry laboratory (the chemical laboratory of the Imperial Institute of Experimental Medicine, the first research medical-biological centre) Ziber-Shumova transferred with him. She continued to act as his assistant making significant advances to the biological studies. When Nencki died in 1901, Ziber-Shumova headed the Laboratory where she continued developing the science proposed by Nencki. In 1912 Ziber-Shumova was appointed a Head of the Chemical Laboratory with the rights of an actual Member of Imperial Institute (equal to a full Professor) and devoted 25 years of her life to the development of biochemistry there. Her publications with Nencki and co-workers dealt with the formation of urea in the organism and role of liver in this process; the chemical composition and biological properties of pepsin; the analogy in chemical structure of haemoglobin and chlorophyll; the processes of oxidation, putrefaction, and fermentation of carbohydrates amongst other studies. Nadezhda O. Ziber-Shumova was the first woman Professor of biochemistry in Russia and she made a vital contribution to the formation and development of the science there. Of the early female chemists, she published what

Fig. 16.2 Upper L-R: Julia Lermontova, Louise Hammarström, Vera Popova (née Boganov-skaya), Edith Humphrey (Arthur Sanderson and Sons via Wikimedia for visual identification) and her *cis*-bis(ethylenediamine)dinitrocobalt(III) bromide. Lower, L-R: Rachel Lloyd, Astrid Cleve von Euler (Uppsala University from Wikipedia for identification), Clara Immerwahr *c*. 1890 (Wikipedia, unknown), Charlotte Fitch Roberts (courtesy *Wellesley College Archives, Library & Technology Services*), and Mary Engle Pennington (from Flickr's The Commons via Wikipedia)

is most likely more papers in the chemical literature than any other woman before 1901 [22]. In 1909 Ziber-Shumova donated to the Polish Society of Biological Investigations 50,000 rubles (roubles) for the foundation of Nencki research institution, which was opened in Warsaw in 1918 (the Nencki Institute of Experimental Biology).

The last of the Russian women worthy of inclusion is Vera Yevstafievna Popova (née Bogdanovskaya) (1867–1896) (Fig. 16.2). Born in St. Petersburg, she spent four years from age 11 taking the Bestuzhev Courses, an institution founded in 1878 in St. Petersburg to encourage Russian women to stay in Russia to study. It was the largest and most prominent women's higher education facility in Imperial Russia, and included amongst its distinguished lecturing staff Alexander Borodin [23]. She then worked in the Academy of Sciences and Military Surgical Laboratories for two years prior to doctoral study at the University of Geneva. Her research there was with Karl Gräbe on dibenzylketone and she gained her Ph.D. in 1892. She returned home and taught at the Bestuzhev Courses giving the first course on stereochemistry. It was there that she wrote her first book, a text on basic chemistry and the first by a Russian woman. It gained a favourable widespread reputation [24]. In 1895 she took a marriage of convenience with the older Jacob Popov, director of a military steel plant, and moved to Izhevskii Zavod, a town in the Western Ural Mountains dedicated to weapons manufacture. One condition of the marriage was that Jacob build his wife a laboratory where she could continue her chemical research. Here she returned to the synthesis of the prussic acid analogue methylidenephosphane (P≡C–H), a topic that Gräbe had dissuaded her from for Ph.D. study because of the perceived difficulty. It is an analogue of hydrogen cyanide in which the nitrogen atom replaced by phosphorus. The first communication on its synthesis appeared only in 1950 and its structure established unambiguously a decade later [25]. This attractive derivative is easily self-ignited and combusts in air at even low temperatures. Sadly, Vera Popova's life was ended on May 8, 1896 following an explosion involving PCH in her laboratory [26]. She was just 28 years old.

Despite the significant influence of Russian women in chemistry in Zurich, they were not the only female students there. Rachel Holloway Lloyd (1839–1900) (Fig. 16.2) was a pupil of organic chemist August Viktor Merz, became the first American female chemist to gain a doctorate, and was, in 1891, the first female admitted to the American Chemical Society. Best known for her work on the chemistry and agriculture of sugar beets, she studied at the Harvard Summer School before receiving her doctorate from the University of Zurich in 1887 for her thesis *On the conversion of some of the homologues of benzol-phenol into primary and secondary amines*. More than a century later in 2014, the ACS designated her research and professional contributions to chemistry a National Historic Chemical Landmark at the University of Nebraska-Lincoln. Another woman of note at Zurich was the British inorganic chemist Edith Ellen Humphrey (1875–1978), who carried out pioneering work in co-ordination chemistry as assistant to Alfred Werner. Gaining her Zurich Ph.D. in November 1901. Her siblings were educated to degree level and she attended Camden School for Girls, then from 1891, North London Collegiate School, one of the first girls' schools in the UK to include science in the curriculum. Between 1893 and 1897, Humphrey studied chemistry (and physics) at Bedford College, London, prior to moving to Zurich to study under Werner. She became the first of his students to prepare the first new series of geometrically isomeric cobalt complexes, a class of compounds crucial in Werner's development and proof of co-ordination theory. One of these compounds, the *cis*-bis(ethylenediamine)dinitrocobalt(III) bromide, was the first synthesis of a chiral octahedral cobalt complex [27]. In 1991, the Swiss Committee on Chemistry donated Humphrey's chiral crystals to the Royal Society of Chemistry, where they are held in Burlington House, London. However, later study has cast doubt on the quality of the sample [27], but Humphrey's status as a pioneer woman scientist remains significant. Her doctoral thesis: *On the binding of compounds to metals and on dinitroethylenediaminecobalt salts* was accepted by the University of Zurich in 1901 and she became the first British woman to obtain a doctorate in chemistry.

The first Swedish woman Ph.D. chemist (second among all disciplines) was Astrid M. Cleve von Euler (1875–1968) (Fig. 16.2). She was a Swedish botanist, geologist, chemist and researcher at Uppsala University. Her Ph.D. was awarded in 1898 at Uppsala University for study of the germinating time and the juvenile stage of some Swedish plants. Astrid's father was the noted Per Teodor Cleve who discovered holmium and thulium. After graduating with a bachelor's degree in 1894, she became an assistant chemistry professor at the progressive Stockholm University. While working there she met the German-Swedish biochemist and subsequent (1929) Nobel laureate Hans von Euler-Chelpin. They married in 1902 and she became Astrid Cleve von Euler, had five children prior to the marriage ending in 1912, and taught in Anna Sandström's women teacher's seminary in Stockholm until 1917. She had published 16 papers with her husband on nitrogenous organics after her papers on lanthanum and selenium appeared from Stockholm. She continued her research at the forestry laboratory at Värmland where she was head.

The beginning of the 20th century saw Germany award its first Ph.D. degree to a woman. Clara Immerwahr (1870–1915) (Fig. 16.2) graduated Ph.D. from the University of Breslau in 1900 from study under Richard Abegg. Her thesis was *Contributions to the Solubility of Slightly Soluble Salts of Mercury, Copper, Lead, Cadmium, and Zinc*. Born of Jewish parents, her father a chemist, she converted to Christianity and married like convert Fritz Haber in 1902. He became the 1918 Nobel Laureate awarded for the Haber-Bosch process, who directed the German use of chlorine in WWI. Clara's work after her marriage was that of assisting her husband. She committed suicide in 1914, some say because of her husband's political views.

Postgraduate study in the USA started at Yale University and the first three Ph.D. students graduated there in 1861. Of these, Arthur Williams Wright was its first science graduate with a degree in physics. Perhaps surprisingly, in 1876, Edward Alexander Bouchet became the first African American gain a Ph.D. (physics, calculus, chemistry, and mineralogy; awarded in physics). By comparison, Marie Maynard Daly (1921–2003), a biochemist, was the first African-American woman to earn a Ph.D. in chemistry, but only in 1947 and from Columbia University. The first seven women Yale Ph.D. students graduated in 1894. Despite this, and because graduate study is (and was) specialised, it was relatively easy for the professors then to avoid teaching female graduate students if they so wished. Admitting women as graduate students, a Yale professor assured the *Times* in March 1892, would not "commit the members of either the academic or scientific departments the university to coeducation in any sense". Of the women pioneers, Charlotte Fitch Roberts (1859–1917) was a chemist. She received her bachelor's degree in 1880 from Wellesley College (near Boston), which made her a graduate assistant in 1881 and an instructor in 1882. She spent the year of 1885–1886 the equivalent of a postdoctoral fellow in Cambridge, England, working with Sir James Dewar. In 1896, she published her book *The Development and Present Aspects of Stereochemistry* [28] with her Yale professor Frank Gooch calling it *he clearest exposition of which we have knowledge of the principles and conditions of stereochemistry*" adding, *there is nothing in English that covers similar ground so broadly and so lucidly.* Roberts became a full professor at Wellesley in 1896, chair of her department, and eventually, a fellow of the American Association for the Advancement of Science.

Charlotte Roberts was followed a year later in 1895 by Mary Engle Pennington (1872–1952) (Fig. 16.2), an analytical chemist, who earned her Ph.D. under Edgar Fahs Smith at the University of Pennsylvania. Pennington's postgraduate work led her to the field of bacteriological chemistry and ultimately to refrigeration engineering. Her work there earned us our confidence in the safe handling, storage, and transportation of foods.

Naturally, there were other women pioneers who did not graduate with a Ph.D. degree but these are too numerus to outline here. The reader may wish to explore: Agnes Luise Wilhelmine Pockels (1862–1935) [29], a German whose work was fundamental in establishing surface science; Josephine Silone Yates (1852–1912), the first African American the first woman to hold a full professorship at any US college or university. Laura Alberta Linton (1853–1915) [30, 31] who, as an undergraduate, analysed mineral specimens collected by her professors that they named *Lintonite*, a rare zeolite mineral; Ellen Gleditsch (1879–1968) [32]. Norway's second female professor and a radiochemist and second woman to be elected to Oslo's Academy of Science in 1917.

References

1. Offereins M (2011) Maria the Jewess. In: Apotheker J, Sarkadi LS (eds) European women in chemistry. Wiley-VCH Verlag GmbH & Co., Weinheim, pp 1–3
2. Wellington J, Bathmaker AM, Hunt C, McCullough G, Sikes P (2005) Succeeding with your doctorate. Sage Publications, London
3. Noble KA (2001) Changing doctoral degrees: an international perspective. Taylor and Francis, Bristol, PA
4. Wikipedia. Timeline of women's education. https://en.wikipedia.org/wiki/Timeline_of_women%27s_education. Accessed 13 Feb 2018
5. Marinnez V (2017) Beyond inspiring: history's female Ph.D. pioneers. A bit of history: exploring historical thoughts and themes, a bit at a time. https://abitofhistoryblog.wordpress.com/2017/09/27/beyond-inspiring-historys-female-phd-pioneers/. Accessed 23 Feb 2018
6. Jansen SL (2015) Juliana Morell—yet another tenth muse. The monstrous regiment of women—a women's history daybook. http://www.monstrousregimentofwomen.com/2015/08/juliana-morell-yet-another-tenth-muse.html. Accessed 15 Feb 2018
7. Thieling S (1995) Elena Lucrezia Cornaro Piscopia. Biographies of Women Mathematicians. https://www.agnesscott.edu/lriddle/women/piscopia.htm. Accessed 28 June 2018
8. Kölbl-Ebert M (2009) How to find water: the state of the art in the early seventeenth century, deduced from writings of Martine de Bertereau (1632 and 1640). Earth Sci Hist 28(2):204–218
9. Offereins M, Strohmeier R (2011) Marie Meurdrac. In: Apotheker J, Sarkadi LS (eds) European women in chemistry. Wiley-VCH Verlag GmbH & Co., Weinheim, pp 13–14
10. Bassi-Veratti Collection. Stanford University Libraries. https://bv.stanford.edu/en/about/laura_bassi. Accessed 6 Mar 2018
11. Laidler KJ, Cornish-Bowden A (1997) Elizabeth Fulhame and the discovery of catalysis: 100 years before Buchner. In: Cornish-Bowden A (ed) New beer in an old bottle: Eduard Buchner and the growth of biochemical knowledge. Universitat de València, Valencia, Spain, pp 123–126
12. Morse EJ (2004) Marcet, Jane Haldimand (1769–1858). In: Oxford Dictionary of National Biography. Oxford University Press, Oxford
13. Rosenfeld L (2001) The chemical work of Alexander and Jane Marcet. Clin Chem 47:784–792
14. Marcet J (1817) Conversations on chemistry, 5th ed. Longman, Hurst, Rees, Orme, & Brown, London
15. Eugster CH (1983) Hundred fifty years of chemistry at the University of Zurich. Chimia 37(6–7):194–237

References

16. Forsius A (2000) Lydia Maria Sesemann (1845–1925)—the first Finnish woman to trained as a doctor. http://www.saunalahti.fi/arnoldus/l_sesema.html. Accessed 9 Mar 2018
17. The matrekeledition files are available for download at http://www.matrikel.uzh.ch/. Accessed 23 Mar 2018
18. Creese MRS (2015) Ladies in the laboratory IV: Imperial Russia's women in science, 1800–1900. Rowman & Littlefield, Lanham, Maryland
19. Lermontoff J (1876) Ueber die Darstellung von Trimethylenbromid. Ann Chem 182(3):358–362
20. Lermontoff J (1879) Ueber die Einwirkung des tertiären Butyljodurs auf Isobutylen bei Gegenwart von Metalloxyden. Ann Chem 196(1):116–122
21. Halton B (2018) Aleksandr Mikhailovich Butlerov (1828–1886) and the Cradle of Russian Organic Chemistry. Chem N Z 82:46–52
22. Creese MRS (1998) Early women chemists in Russia: Anna Volkova, Iuliia Lermontova, and Nadezhda Ziber-Shumova. Bull Hist Chem 21:19–24
23. Halton B (2014) Alexander Porfirevich Borodin (1834–1887). Chem N Z 78(1):41–47
24. Rulev AY, Voronkov MG (2013) Women in chemistry: a life devoted to science. New J Chem 37:3826–3832
25. Gier TE (1961) HCP, a unique phosphorus compound. J Am Chem Soc 83:1769–1770
26. Eleanor S, Elder ES, Lazzerini SD (1979) The deadly outcome of chance-Vera Estaf'evna Bogdanovskaia. J Chem Educ 56:251–252
27. Ernst KH, Wild FRWP, Blacque O, Berke H (2011) Alfred Werner's coordination chemistry: New insights from old samples. Angew Chem Int Edn 50:10780–10787
28. Roberts CF (1896) The development and present aspects of stereochemistry. D. C. Heath & Co., Boston, p 1896
29. Pockels A (1891) Surface tension. Nature 43:437–439
30. Linton LA (1894) On the technical analysis of asphaltum. J Am Chem Soc 16:809–822
31. Linton LA (1896) Technical analysis of asphaltum. No. 2. J Am Chem Soc 18:275–283
32. Lykknes A, Kvittingen L, Børresen AK (2005) Ellen Gleditsch: duty and responsibility in a research and teaching career, 1916–1946. Hist Stud Phys Biol Sci 36:131–188

Chapter 17
William John Young (1878–1942)

William John Young (Fig. 17.1) was born in Withington, Manchester in 1878, the son of William John Bristow Young, a clerk, and his wife Hannah (née Bury). He was educated at the local Hulme Grammar School, where he showed interest in science. On leaving school, the boy enrolled at Owens College in Manchester and graduated with a B.Sc. in 1898 beginning research early in his career. He won the Levinstein and Dalton research exhibitions for the 1899 and 1900 years, respectively, and graduated with his M.Sc. in 1902. On July 30, 1903, he married Janet Taylor at St. Margaret's parish church in Whalley Range, Manchester.

William Young worked in collaboration with (Sir) Arthur Harden at the Lister Institute in London and also whilst at Owens College, as he was appointed Assistant Biochemist from there in 1901. This work led to the major collaborative efforts of Harden and Young that culminated in the establishment of biochemistry as a discipline, and to Harden receiving the 1929 Nobel Prize in Chemistry. William's work concerned the mechanism of fermenting enzymes in yeast extract, and extended the work of Eduard Buchner [1] on cell-free alcoholic fermentation. A 1901 paper (with Roland) [2] on the auto-fermentation and liquefaction of yeast cites the assistance of Young after his appointment as assistant to Harden but prior to his arrival at the Lister in early 1902. This acknowledgement has to be to William's Manchester experiments. It was these that began their fruitful collaboration that lasted some 10 years.

The great achievement of Harden and Young was the discovery of phosphorylation and dephosphorylation that forms the basis of carbohydrate metabolism. They elucidated the manner in which this process occurs from discovering that a coenzyme existed, and they suggested that its function was to take up and pass on the phosphate radical. Their evidence of phosphorylation was indisputable as they isolated two intermediates in which phosphate was bound to hexose: 1:6-hexose diphosphate (**1**) and 6-hexose monophosphate (**2**) (Fig. 17.1). The diphosphate, coined the Harden-Young ester was later shown to be fructose-1,6-diphosphate (**1**) and the first chemical

---

A previous version of this chapter was published as Halton B (2014) Sir Arthur Harden, FRS (1865–1940) & William John Young (1878–1942). Chem N Z 78:169–174.

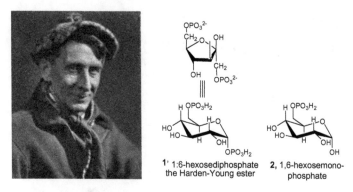

Fig. 17.1 Left: William John Young (courtesy the University of Melbourne Archives), centre the Harden-Young ester, hexose diphosphate (**1**), and right: hexose monophosphate (**2**)

intermediate discovered in fermentation [3]. Its discovery led to the ultimate description of fermentation in terms of molecular intermediates:

1. $2C_6H_{12}O_6 + 2Na_2HPO_4 = C_6H_{10}O_4(PO_4Na_2) + 2H_2O + CO_2 + 2C_2H_6O$
2. $C_6H_{10}O_4(PO_4Na_2)2 + H_2O = C_6H_{12}O_6 + 2Na_2HPO_4$

In 1910, William Young received his D.Sc. from the University of London and the pair laid the foundations of carbohydrate biochemistry.

During the fruitful collaboration with Harden in London, the position of Biochemist at the Australian Institute of Tropical Medicine in Townsville became available and Young was appointed by a selection committee chaired in London by Dr. C. J. Martin, the then Director of the Lister Institute. With his wife and young daughter Sylvia, Martin left London in late 1912 for Townsville [4–7]. After seven years there he moved to a lectureship at the University of Melbourne and rose to become the Foundation Professor of Biochemistry in 1938.

Proposals for the Tropical Medicine Institute in Townsville were adopted in 1908 and Austrian Dr. Anton Breinl was appointed its inaugural Director from the Institute of Tropical Medicine in Liverpool. He arrived in Queensland on January 1, 1910, holding the reputation as one of the most promising medical scientists working in Britain; he had discovered an organic arsenical cure for sleeping sickness in 1904. Subsequently Ehrlich exploited he discovery to produce salvarsan (arsphenamine) for the treatment of syphilis. From 1910 until 1912, Breinl and his laboratory assistant (Fielding) were housed in a three-roomed building in the grounds of Townsville Hospital. It was only in 1912 when John Nicol, Henry Priestley and William Young were appointed to the staff that laboratory science was initiated in Northern Queensland allowing Townsville to become the birthplace of Australian biochemistry [8–10].

It is said the Young family had a happy time in Townsville [4, 5]. William's work there, both alone and with Breinl was dominated by the metabolism of "whites living in the tropics" and these studies led to four joint papers with Breinl, most notably that in the *Annals of Tropical Medicine and Parasitology* in 1920 [11]. He had a further four publications on the subject alone and another five on diverse topics

[12]. The success of the Institute [8–10] was such that, by 1920, it had adequately demonstrated that the tropical regions were habitable by a "working white race", as specific conditions such as leptospirosis (or Weil's disease) had been understood. It had served its purpose and was disbanded in 1921 to be subsumed into the newly founded Australian Government Commonwealth Department of Health.

In 1920, W. A. Osborne, Professor of Physiology at the University of Melbourne, secured William Young's appointment as lecturer in biochemistry and he moved there to take charge of the subject. Soon after arriving, he developed an interest in the applied biochemistry of food preservation and transport and was promoted to Associate Professor of Biochemistry in 1924; most likely, he was the first academic in Australia to take a continuing interest in what is now recognised as food science and technology [13, 14].

The year 1926 saw (Sir) Albert Rivett persuade Melbourne University to release Young to help the Council for Scientific and Industrial Research (CSIR) (with which he had worked closely) to establish laboratory and field studies into the biochemical problems of the cold storage of food. Despite Rivett's hopes of recruiting William to the CSIR, instead he returned to his teaching duties in 1928, which included lecturing to agricultural, dental, medical and science students. Nonetheless, he remained a consultant to the CSIR and his recommendations led to the formation of the section of food preservation and transport in 1931. It was only in 1938 that William Young was appointed the inaugural Professor of Biochemistry at the University of Melbourne. By then he had solved problems associated with the conveyance of chilled meat and the preservation of citrus fruits. Some of the methods that came from his studies are still used in the marketing of ripening bananas [15]. The Brisbane abattoir laboratory to study meat chilling had opened in 1932 [16]. With the co-operation of Professors Bagster and Goddard of the University of Queensland, Young guided and planned in detail investigations on meat, fish, bananas and citrus, and provided room for the laboratory work to be performed in his own department. He was the first to train food technologists (then not known as such) and he assisted C. P. Callister in his researches to become probably the first (1931) Australian D.Sc. in food technology [17]. It was Callister who had developed Vegemite from brewer's yeast and marketed it in Australia in 1923.

William John Young was an active member of the Society of Chemical Industry of Victoria giving lectures and publishing various aspects of his work. Examples include *The Science of Fruit Preservation, Refrigeration of Beef, Preparation and Transport of Australian Chilled Meat*; *The ripening and transport of bananas in Australia*, and *An Experimental Shipment of Navel Oranges to Canada*. He had collaborated with Osborne whilst at the Townsville Institute [16] and he co-authored the text *Elementary Practical Biochemistry* with him through the first five of its six editions; it was published in Melbourne by W. Ramsay.

He was also an enthusiastic bushwalker and member of the Melbourne Wallaby Club, holding the presidency during 1925–26. His contemporaries spoke of him as an indefatigable worker and a charming companion. He was gentle and modest with an uncommon sense of social involvement. He supported his profession holding the presidency of the Victorian Branch of the Australian Chemical Institute over 1938–1939. William John Young died on 14 May 1942 in East Melbourne from a perforated gastric ulcer.

References

1. Buchner E (1897) Alkoholische Gährung ohne Hefezellen. Ber Dtsch Chem Ges 30:117–124
2. Harden A, Rowland S (1901) Autofermentation and liquefaction of pressed yeast. J Chem Soc Trans 79:1227–1235
3. Korman EF (1974) The discovery of fructose-1,6-diphosphate (the harden-young ester) in the molecularization of fermentation and of bioenergetics. Mol Cell Bioc 5:65–68
4. Smedley-MacLean I (1943) Obituary notice— William John Young. Biochem J 37:165–166
5. Slater EC (2004) Cozymase, coenzyme I (II), D(T)PN and NAD(P). IUBMB Life 56(5):289–290
6. Marginson M (1990) Young, William John (1878–1942). Australian Dictionary of Biography, Australian National University. http://adb.anu.edu.au/biography/young-william-john-9220/text16291. Accessed 7 Feb 2014
7. Maxwell I, Hopkins FG, Martin CJ (1943) Obituary notice—William John Young. J Chem Soc 44–45
8. Douglas RA (1977) Dr. Anton Breinl and the Australian Institute of Tropical Medicine. Part 1. Med J Austral 1:713–716
9. Douglas RA (1977) Dr Anton Breinl and the Australian Institute of Tropical Medicine. Part 2. Med J Austral 1:748–751
10. Douglas RA (1977) Dr. Anton Breinl and the Australian Institute of Tropical Medicine. Part 3. Med J Austral 1:784–790
11. Breinl A, Young WJ (1920) Tropical Australia and its settlement. Annals Trop Med Parasitol 13:351–412
12. For a list of his 14 publications, see: Publications from the Australian Institute for Tropical Medicine: 1903–1930. http://www.tropicalhealthsolutions.com/node/33. Accessed 4 Apr 2014
13. Vickery JR (1986) W. J. Young, a pioneer in food science in Australia. Food Technol Austral 38:158–159
14. Bastian JM, McBean DG, Smith MB (1979) 50 years of food research. CSIRO, Melbourne, pp 9–10
15. The Charleville Times, 12 February 1932, p 1
16. Farrer KTH (2011) Birth of a profession: a history of the Australian Institute of Food Science and Technology. Australian Institute of Food Science and Technology, Alexandria, NSW, p 1
17. Young WJ, Breinl A, Harris JJ, Osborne WA, Langley JN (1920) Effect of exercise and humid heat upon pulse rate, blood pressure, body temperature, and blood concentration. Proc Roy Soc (Lond) B Biol Sci 91:111–126

Printed in the USA
CPSIA information can be obtained
at www.ICGtesting.com
LVHW010936190324
774517LV00003BA/290

9 783030 164027